The Herbarium
HANDBOOK

Sharing best practice from across the globe

Edited by
**Nina M.J. Davies, Clare Drinkell
& Timothy M.A. Utteridge**

Kew Publishing
Royal Botanic Gardens, Kew

First published in 2023 by
Royal Botanic Gardens, Kew, Richmond, Surrey, TW9 3AB, UK. www.kew.org

ISBN 978 1 84246 769 5
eISBN 978 1 84246 770 1

British Library Cataloguing in Publication Data
A catalogue record for this book is available from the British Library.

Design: Nicola Thompson, Culver Design
Production Manager: Georgie Hills
Proofreading: Matthew Seal
Index: Sharon Whitehead

Printed in Italy by L.E.G.O. S.p.A.

For information or to purchase all Kew titles please visit shop.kew.org/kewbooksonline or
email publishing@kew.org

Kew's mission is to understand and protect plants and fungi, for the wellbeing of people
and the future of all life on Earth.

Kew receives approximately one third of its funding from Government through the
Department for Environment, Food and Rural Affairs (Defra). All other funding needed
to support Kew's vital work comes from members, foundations, donors and commercial
activities, including book sales.

CONTENTS

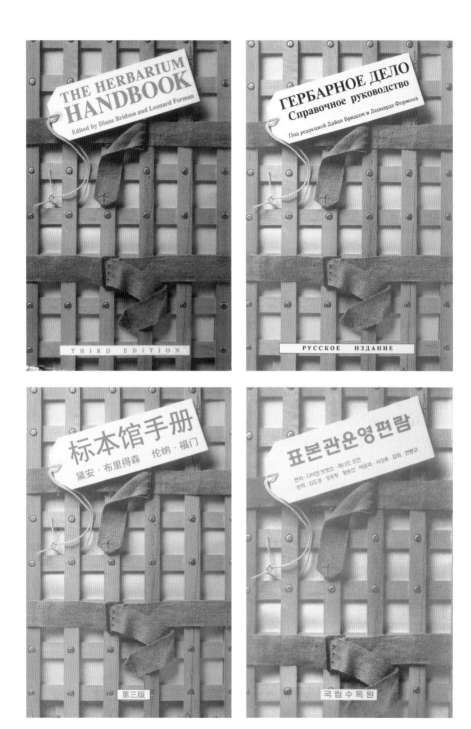

The *Herbarium Handbook* 1989–1999 in English, Chinese, Russian and Korean

PREFACE

Nina Davies, Clare Drinkell and Timothy Utteridge

History

The Herbarium Handbook was first published by Kew in 1989 following its popular International Diploma Course in Herbarium Techniques. Subsequently it has been reprinted and translated into Chinese, Korean and Russian; by 1999 this was on the third edition of the book. The book is still widely used, and for us Editors it was the first manual that we were given when starting our careers at the Royal Botanic Gardens, Kew. Over the years, the book has served us well, and, like an old friend, is still a valuable resource for some of the fundamental aspects of herbarium curation. The latest version updated was nearly a quarter of a century ago, and in recent decades many parts of the book, such as how to copy a floppy disk drive(!) have drifted away from technological advances – databasing, imaging and 'big data' are driving future directions of herbarium curation. Now, using more modern printing techniques, digital photography and drawing experiences from our friends and colleagues in herbaria across the world, we have tried to update the book for modern times. The very first edition was compiled and edited by two Kew botanists, Leonard Forman and Diane Bridson, two very experienced botanists and the essence of the original *Herbarium Handbook* is something we hope to have captured in this new book.

We aimed to edit a book with many different voices, rather than as before, a manual of best practice conducted in the Kew Herbarium – one of the largest in the world with dedicated teams of staff addressing all aspects of the curation, management, care and research. The resources available to the world's 3500+ herbaria are unevenly distributed, but it is the smaller, regional herbaria that are well known to be disproportionately representative of the local flora (Monfils *et al.* 2020), with coverage much better than that of Kew, or any other of the large global herbaria. We have therefore consulted widely, drawing on the expertise of authors outside Kew, and setting out best practice applicable in herbaria of any size or location. All of us work in institutions of different sizes, with different histories, strategies and challenges, but the fundamentals, to collect and preserve the world's plants in herbaria for now and future generations, remain the same.

September 2023

EX NASIONALE } HERBARIUM, PRETORIA,   Suid-Afrika.
NATIONAL }   South Africa.

Prov. _Bechuanaland_   Dist. _____

Ornithogalum seineri (Engl. + K₂)
Obern. c. n. (Bothalia 1964)
syn. Urginea langii Brem.
(Bulbine seineri Engl. + K₂)
syn. Ornithogalum wilmaniae Leight.
from Derdepoort just over the
Transvaal border. Flowered
at Prinshof, Pretoria, 30-10-'59

Pre. No. _____   Legit _L. E. Codd_

Alt. _____ Anno. _____   No. _8875_

ACKNOWLEDGEMENTS

Nina Davies, Clare Drinkell and Timothy Utteridge

This volume is the first comprehensive revision and update of *The Herbarium Handbook* that was first published by Kew in 1989. Although not a 'like-for-like' change from the previous content, we would like to acknowledge the editors Diane Bridson and Leonard Forman, as well as all the contributors, who compiled and produced the first handbook: without them we would not have a project to update.

Gina Fullerlove, then Head of Kew Publishing, was instrumental in pushing for a new edition, and we would like to thank her for support and enthusiasm for the project from its inception. Alan Paton, Mark Nesbitt and Lauren Gardiner drafted an initial concept for the new edition, and we would like to thank them for getting the ball rolling. In addition to Gina, we would like to thank the Kew Publishing Team, especially Lydia White and Georgie Hills, for their support and guidance throughout the process of completing the handbook, while Paul Little took many of the photographs used throughout the book and we thank him for his patience.

The writing and production of this book has involved many people from Kew and herbaria across the world. We would like to warmly thank all the contributors for submitting all their content in a timely fashion; we hope we have reflected different voices and ways of working from herbaria across the world and are extremely grateful for your patience and understanding from submission to publication. We would like to thank all those involved for adaptability during the production and editing process, which started during the COVID pandemic – we acknowledge everyone was struggling to understand the new ways of working online but we are now better connected than before, and hopefully this handbook will do a small part to further reinforce the connections between the global herbarium community.

We would like to especially acknowledge the enthusiastic and knowledgeable input of Alan Paton, who has been supportive of the project throughout, especially with additional financial support to help support the project. His experience and wide-ranging knowledge of the herbaria, their history and practices, was helpful in guiding and advising on content during the development and then writing. A very special thanks goes to Noah Hearne for helping us to check much of the content, photographs and to tie up loose ends during the project. Our colleagues at CAS would like to pay their gratitude and respects to Dr. Nathalie Nagalingum (1975–2022); her strong belief in outreach and education helped to initiate and organise their contribution to the handbook.

Finally, we would like to thank everyone who works tirelessly to maintain and enhance the world's herbaria, keeping them as the essential resource for all plant diversity research in the past, present and future.

INTRODUCTION

PURPOSE OF THE BOOK
Nina Davies and Clare Drinkell

Editions 1 to 3 of *The Herbarium Handbook* have been an important reference for herbarium collections care and management since first published in 1989 (Forman & Bridson 1989). The handbook brought together practical guidance covering aspects of technical herbarium work written exclusively by staff within the Kew Herbarium. This included procedures on practical herbarium techniques such as the preparation, preservation and organisation of collections; fieldwork and collecting; and the management of the building environment. Herbaria worldwide have been able to adapt the guidance from *The Herbarium Handbook* to their specific requirements.

A collaboration
This book has taken on a new approach and is a complete renewal of the previous editions of *The Herbarium Handbook*, incorporating best practice methods from a wide range of herbaria – small to large, temperate to tropical. We see it as a celebration of collective knowledge. The chapters are arranged in a logical structure from field collection, to herbaria, with useful anecdotes, examples and statistics from Kew staff, but crucially partners and collaborators from around the world who play a key role in contributing to the handbook.

New sections cover biosecurity, digitisation and herbaria in the wider context of public engagement and outreach. This book, which also includes photographs and illustrations, is a richly illustrated reference tool offering contemporary herbarium management from wide-ranging collaborators and informed by best practice from the Kew Herbarium. The updated *Herbarium Handbook* will continue to be an important reference book, to help with training future generations of staff, interns and volunteers, share ideas on techniques and workflows and advise on time-saving and cost-effective processes.

A BRIEF HISTORY OF HERBARIA

Craig Brough and Timothy Utteridge

An herbarium is a collection of preserved plant specimens, with each specimen mounted on a sheet of paper and kept in an appropriate repository to ensure their accessibility and safety. They are the fundamental resource for botany (Funk 2003a). Globally, there are thousands of herbaria, each with many thousands of specimens that have accumulated over centuries of botanical endeavour.

Luca Ghini (1490–1556), a physician and lecturer at the University of Bologna, is credited with compiling the first herbarium (Thiers 2020). Ghini also established the first course of study in plants for their own sake rather than as a component of pharmacognosy. To enable his students to study the morphology of plants even in winter, Ghini dried living plants by pressing them between sheets of paper and then gluing them into the blank sheets of a book. This had the advantage over the illustrated books of the time, since the reproduction of illustrations was not sophisticated enough to enable plants to be precisely identified but, with dried specimens, they could be.

Subsequent herbaria were initially bound as books in a similar way. Ghini's herbarium is not known to have survived, but those of his students have.

1 Carl Linnaeus (1707–1778), often known as 'the father of taxonomy' who formalised the binomial system of nomenclature. **2** The Sir Hans Sloane Herbarium contains plant specimens collected on Sloane's voyage to Jamaica (1687–1689); the specimens were mounted in seven bound volumes, which have been preserved intact. **3** Royal Horticulture Society, Wisley. Opened in 2021, the new RHS Herbarium holds the UK's largest collection of cultivated plants, with more than 86,000 dried plant specimens.

With the expansion of European trading expeditions, opportunity arose to collect plants new to European science, particularly those with a commercial value. William Dampier (1651–1715) is considered to be the first traveller charged with collecting herbarium specimens. Most were lost to shipwreck, but those that did survive (from Brazil, Australia and New Guinea) passed in time to Oxford University. During the period 1500–1950 imperial networks enabled the flow of plant specimens from colonised regions to the academic centres of Europe, as part of a wider programme of the classification and exploitation of biological resources.

The herbarium as we know it today, loose sheets of specimens housed together in a cabinet, can be traced to Carl Linnaeus (1707–1778). This enabled new collections to be incorporated with older ones while preserving the relationships of similar species. The Linnean herbarium, amounting to about 14,000 specimens, was purchased by wealthy English botanist James Edward Smith and transferred to London, where it resides at the Linnean Society (Savage 1945).

All herbaria are listed on 'Index Herbariorum', an online resource hosted by the New York Botanical Garden (Thiers, continuously updated). Index Herbariorum was established by the International Association for Plant Taxonomy (IAPT) in 1935, and the first lists were produced as published editions, but it has been online since 1997. It is the fundamental resource to discover herbaria from around the world: each listing has contact information, key information about the collections, staff details and lists of important collectors.

As of 2021, Index Herbariorum lists 3522 active herbaria globally with nearly 400 million specimens (Thiers 2022). Each herbarium will have a distinct combination of specimens, coverage and expertise. Regional herbaria will have more focused, and unique, collections from a particular geographical area, whereas international herbaria have a more general coverage from across the globe.

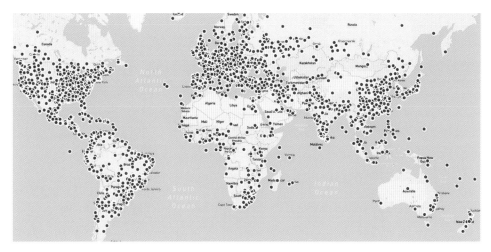

Global distribution of active herbaria (detail from Thiers, continuously updated).

WHY DO WE NEED HERBARIA?

Manuel Luján and Ana Rita G. Simões

Herbaria are repositories of plant specimens and their ancillary information which are maintained as physical records of the world's plant diversity. Technological advances applied to natural history collections (e.g., DNA, digitisation) have unlocked valuable biological information, beyond the physical specimen. Hence, herbaria are fundamental for virtually all branches of plant science.

Natural history collections, and their associated data, provide the foundation for our understanding of the world's living organisms including their morphological, genetic, physiological, behavioural, and ecological variation. Herbaria were originally established as reference collections of plants with medicinal and economic value, as well as to support scientific endeavours such as plant identification, floristic inventories and new species descriptions. Herbaria are paramount sources of information and remain essential to botanical research: new species are routinely described, nomenclature is continuously updated, and documentation of the world's plants is increasingly relevant as we face an era of vast biodiversity loss (Heberling & Isaac 2017). Maintaining and caring for herbaria, both the physical specimens and digital resources, is critical for their current and future use. Moreover, as unparalleled records of plant diversity, herbarium specimens provide critical evidence to address the impact of anthropogenic activities on the environment including global climate change and all its disruptive consequences (Bakker *et al.* 2020).

Herbarium specimens as physical records of diversity

The world's 3522 herbaria contain over 397 million specimens, which have been gathered and curated through the efforts of thousands of botanists for over four hundred years (Thiers 2022). Plant specimens are dried, pressed, mounted and incorporated into herbaria where they are maintained as tangible records of the plant diversity that exists or has existed in determined areas and times. Herbarium specimens are essential for comparative studies in taxonomy, systematics, conservation biology, ethnobotany, paleobiology, as well as being valuable resources for teaching students at all levels and educating the general public (Funk 2018). While photographs and other types of information can be recorded from a living plant, keeping a physical record of an individual or a population of plants unequivocally preserves all its characteristics, at all scales of dimensions, from its overall habit to its tissues and genomic information. Hence, the fundamental value of herbaria relies on the adequate preservation of these plant specimens, and all their ancillary information.

The most important assets of herbaria are the plant specimens they house. Maintaining these specimens in adequate conditions ensures verification and repeatability, which are essential to the scientific process. In this context, advancing plant systematic and taxonomic research is only possible if scientists can reassess appropriately preserved specimens.

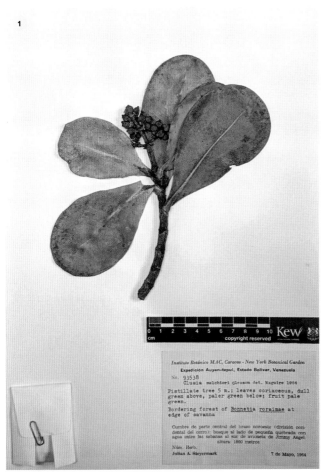

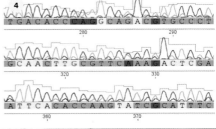

Common uses of herbarium specimens in scientific research. **1** Herbarium specimen. **2** Scanning electron microscope image of pollen grains. **3** Representation of a morphometric analysis based of leaf measurements. **4** Fragment of DNA sequence obtained from leaf tissue. **5** Species distribution map based on locality data obtained from specimen labels.

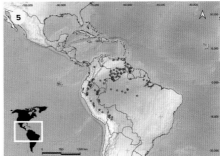

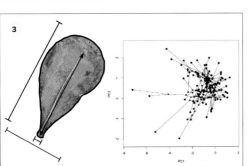

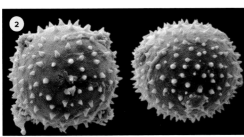

Virtual collections

Modern technological and analytical methods utilise specimens in non-traditional and unanticipated ways (Funk 2018; Heberling & Isaac 2017). The digitisation of collections, for instance, has been an extraordinary breakthrough in the use of herbaria, making images of plant specimens and related information, such as locality data and phenology, readily available to the scientific community worldwide. Integration of specimen data into freely available and searchable databases (e.g., iDigBio, GBIF, Atlas of Living Australia, Canadian Museum of Nature, Reflora Virtual Herbarium) has opened new avenues to exploring these virtual collections. As a result of these initiatives, images from herbarium collections can be used for studies which benefit from analysing large numbers of specimens, such as species distribution modelling and morphometric studies. Furthermore, the integration of digitised collections in machine learning processes offers exciting opportunities to exploit the vast amount of information contained in herbarium specimens.

Digitisation of natural history collections makes them accessible to scientists around the world, which is particularly important to researchers and students based in countries where institutional support for travelling internationally is deficient. Moreover, digitised specimens are being used to engage the public in biodiversity research, particularly on initiatives where citizen scientists are contributing on research projects using online platforms such as Digivol, Les Herbonautes, Zooniverse.

Extended specimen (associated data)

The concept of the extended specimen (Webster 2017) captures the multitude of data types that can be extracted and recorded from a single specimen, and places specimens as essential hubs of information from which new areas of biological research may expand (Lendemer *et al.* 2020).

Historically, herbarium specimens have been used mostly to study the morphological diversity of plants across space and time. Using modern molecular techniques, herbarium specimens are currently essential for exploring plant genetic diversity, allowing better understanding of patterns and modes of evolution. Furthermore, herbarium specimens can be used to analyse biochemical characteristics such as carbon isotopes, nitrogen and phosphorous contents in various tissues, allowing comparative studies of plant physiology. Pollen grains and spores contained within herbarium specimens are critical sources of material for palynological studies that help us reconstruct the composition of extinct plant communities. In addition to the dried and pressed plant specimen, plant materials such as seeds, fruits, and wood sections, are frequently collected and maintained in herbaria, and they are fundamental sources of evidence in plant micromorphology and anatomy studies.

CATEGORY	EXAMPLES
Taxonomy	Species descriptions
	Floristic inventories
	Taxonomic revisions
	Field guides
	Identification keys
Tissue sampling	DNA (Genomics)
	Phytochemistry, stable isotopes
	Pollen
	Seeds
	Anatomy
Images	Morphometric studies
	Automated plant identification (AI)
	Wider access to collections
Geographic information	Species distribution modelling, Geospatial analyses
	Rapid Conservation Assessments
	Important Plant Areas
Historical/economic value	Biography of collectors
	Economic use of plants
	History of the exploration of a region
	Documenting field campaigns
Temporal frame	Phenology
	Anthropogenic effects, climate change

Examples of uses of herbarium specimens.

HERBARIUM HIGHLIGHT:
ROYAL BOTANIC GARDENS, KEW (K)
Timothy Utteridge

The Herbarium at The Royal Botanic Gardens, Kew was founded based on the collections of William Hooker and George Bentham. The Herbarium is arranged by the Angiosperm Phylogeny Group (APG) classification of plant families and genera across several purpose-built wings.

Collections and collectors

The Kew Herbarium (K) was officially founded in 1852 though the genesis of the collection was the personal herbarium of the first Director, William Jackson Hooker (1785–1865; Director of Kew in 1841; see Desmond 1995). Today, the collection is representative of global plant diversity and includes c. 7 million vascular plant specimens with coverage of around 95% of vascular plant genera. A separate Fungarium holds approximately 1.25 million fungal specimens.

K holds approximately 360,000 types and all are digitised and available online; the remaining Herbarium collections are currently being digitised.

Environment

There are five Wings at Kew – only two are air conditioned, the other three have the specimens stored in wooden cupboards which buffer the environmental conditions. Fire detection systems are present throughout all the Wings. The large atria and flared window casings allow natural light into the Wings – perfect for studying specimens.

Important early accessions
(compiled by David Goyder)

1854 – Herbarium and library of George Bentham given to Kew

1858 – Indian collections of Griffith, Falconer and Helfer acquired from the East India Company together with Roxburgh drawings

1862 – Borrer's British plants; Cunningham's collections from Australia

1863 – Grant's East African plants collected on Speke & Grant expedition to discover the source of the Nile

1865 – Burchell's collection from St Helena, South Africa and South America presented; Lindley's orchid herbarium purchased

1866 – WJ Hooker's herbarium (£5000), library (£1000) and correspondence (£1000) purchased by the Government after his death and formally incorporated into the collections at Kew

1877 – Schweinfurth's tropical African plants; Indian herbarium of CB Clarke

1878 – Dalzell's Indian herbarium

1880 – Herbaria of Schimper (N.E. Africa), Bishop Goodenough and W Munro.

1881 – Watson's British plants

1882 – Baron's herbarium (Madagascar); herbarium of Botanical Record Club; Leighton's herbarium

1913 – Herbarium of the Honourable East India Company (the "Wallich" herbarium) transferred from Linnean Society of London

Scope

Current activities are focused on the wet tropics, including Amazonia, West Africa, Madagascar and South-East Asia (Thailand to New Guinea). Around 20,000 specimens are added to the collection each year, a quarter of which are collected by Kew staff with partners from around the world, the remainder being sent from other herbaria worldwide. The Herbarium has a long history of writing and supporting complex Flora projects, such as Flora of Tropical East Africa (FTEA).

Research

The Herbarium supports a wide range of research at Kew, especially taxonomy and systematics of tropical plant groups, phylogenetics, conservation assessments and supporting seed banking and other conservation initiatives, including 'Tropical Important Plant Areas'.

Visits and loans

Specimens are sent on loan (except the East India Company Herbarium K-W) and c. 450 researchers visit the Herbarium each year from other institutions to study our collections for their projects. Visits are requested through herbarium@kew.org and www.kew.org. Sampling of specimens, e.g., for DNA and phylogenomics, is allowed through Material Transfer Agreements.

The Herbarium at Kew. **1** Wing C, the first purpose-built herbarium wing completed in 1877. **2** Wing B, with a similar Victorian style to Wing C, completed in 1902. **3** Wing A dates from 1932. **4** Wing E is the most recent wing and was opened in 2009.

COLLECTING FOR THE HERBARIUM

INTRODUCTION TO COLLECTING
Alan Paton

This section describes the role of the herbarium specimen as verifiable evidence to support research and considers factors which influence what should be collected.

Why collect? The specimen as a voucher

Herbaria are used in a large variety of ways and, fundamental to all those uses, herbarium specimens record what plants occurred where and when, enabling the specimen to be viewed in spatial, taxonomic, evolutionary and historical contexts. Herbarium specimens thus provide auditable data on species occurrence through time, enabling modelling of future distribution. Their digitisation facilitates the linkage of an occurrence to other information derived from the specimen itself, such as DNA sequences, anatomy or images or to information about the locality, climate, literature or historical events, broadening the potential users of the collection. The herbarium specimen thus provides a voucher supporting research which can be verified through time. We collect herbarium specimens to record and facilitate research, to provide opportunities for study by future research techniques and to connect people to plant diversity in education and outreach activities.

Targeting collection

Data from collections are clearly a vital source of information for understanding the world around us, but that in itself does not explain what we should collect. Collections need to be gathered, looked after over time and information needs to be disseminated in an effective way. Collecting activities should be targeted so that the resources available can be focused on ensuring the material collected and the information gathered can be used to add value to current knowledge and provide opportunities for further research and use. This focus will be determined by the broader long-term goals of the institution, existing collection strengths and weaknesses, the needs of intended users of the collection and how they can be best served.

Collection constraints

National law on access to land or genetic resources may impose limits and conditions on collecting. Collectors should also collect responsibly, collecting only what is required to meet the aims of the herbarium and avoid over-collection that can endanger the natural population of the plants being collected.

Collecting and pressing fresh material in the Tamrau Mountains, Indonesian New Guinea.

LEGISLATION
ACCESS AND BENEFIT SHARING LEGISLATION
China Williams

The 1993 Convention on Biological Diversity (CBD) balances objectives to conserve biological diversity and sustainably use its components, with the need to fairly and equitably share benefits from the use of genetic resources with providers. Individual countries' National Legislation sets out how this can be achieved.

The Convention on Biological Diversity (CBD)
There is now a range of legislation, both national and international, that impacts the way in which we collect, use and supply plant specimens. The CBD or 'Biodiversity Convention' aims to conserve biodiversity, promoting its sustainable use and the fair and equitable benefit-sharing arising from the use of genetic resources.

The CBD came into force in 1993, and there are now 196 Parties (countries) who have agreed to translate these objectives into national legislation. A Conference of Parties (COP) takes place every 2 years for discussion on implementation of the CBD.

The Nagoya Protocol
The third objective of the Convention – fair and equitable benefit sharing from the utilisation of genetic resources – is now negotiated through the legally binding 'Nagoya Protocol on access and benefit sharing'. The Nagoya Protocol (NP), which came into force in October 2014, aims to provide more legal certainty to users and providers of genetic resources. Parties' access legislation must be clear on how users obtain prior informed consent (PIC) and mutually agreed terms (MAT), and all Parties must introduce compliance measures (such as checkpoints) to ensure genetic resources utilised have been legally accessed, to monitor utilisation, and to address situations of non-compliance.

It requires users to share benefits from the utilisation of genetic resources with the providers of those resources – with the aim that these benefits (money, capacity, and technology) will ultimately support, and provide an incentive for, the conservation of those same resources. The Nagoya Protocol also applies to the utilisation of Traditional Knowledge associated with genetic resources.

Access and Benefit Sharing Legislation
Many Parties to the CBD and Nagoya Protocol now have national legislation to put it into practice. This will set out procedures that need to be followed and will usually be through a collecting permit or other letter of authority or written agreement with the relevant ministry in the provider country. In addition, or as part of this, you will need to negotiate fair and equitable benefit sharing, which could include monetary benefits (for instance, royalties from commercial application) or non-monetary ones (including providing research results, capacity building, etc.), depending on the project. These PIC and MAT documents can be published by the providing country on the Access and Benefit Sharing Clearing House website as Internationally Recognised Certificate of Compliance (IRCC), which can be referred to by users down the

supply chain to ensure material used has been legally accessed.

You can check the ABS Clearing House to find up-to-date legislation for each of the Parties and also, a list of IRCCs issued by the Parties.

The NP is intended to be implemented in a 'mutually supportive' manner with other relevant ABS instruments – such as the International Treaty on Plant Genetic Resources in Food and Agriculture (ITPGRFA), which incorporates benefit sharing agreements. The international system of Plant Breeders' Rights (PBR) is considered not to conflict with the NP, in that it does not impose restrictions on further use of plant genetic resources other than its propagation for sale.

TIP

The Nagoya Protocol does not replace or negate existing laws, regulations, or agreements on access or benefit sharing that many countries already have in place. These may include legal collecting permits, permissions from landowners, the Convention on International Trade in Endangered Species of Wild Fauna and Flora (CITES) and other export permits, local legislation, rules on working with indigenous communities, permissions to collect endangered or threatened plants, plant health regulations, etc. You still need to comply with these requirements, as well as any additional access and benefit sharing legislation a country has in place to implement the Nagoya Protocol.

Most genetic resources currently being utilised will have been accessed well before 12 October 2014. These will continue to be subject to the terms and conditions under which they were acquired, and these must be respected, just as any other bilateral contract. In addition, many countries are not Parties to the Nagoya Protocol but still have legislation and permitting requirements that need to be followed when collecting plants or animals.

FURTHER READING
CBD Website: www.cbd.int
ABS Clearing House https://absch.cbd.int/

A Yanomami woman, in the northern Brazilian Amazon, is recording traditional knowledge of medicinal plants from the older knowledge holders. Several young indigenous researchers of both genders were trained, and the information was published in an illustrated bilingual book in Yanomami and Portuguese.

CONVENTION ON INTERNATIONAL TRADE OF ENDANGERED SPECIES (CITES)

Sonia Dhanda

The Convention on International Trade in Endangered Species of Wild Fauna and Flora (CITES) aims to protect listed species of plants and animals against overexploitation caused by international trade and to ensure this trade is sustainable. Herbaria have an important role to play in implementing CITES.

Which plants are covered by CITES?
Over 34,000 plant species are listed on CITES. Species are listed on one of three Appendices, I, II and III:

- **Appendix I** lists species threatened with extinction. Trade is permitted only in exceptional circumstances, such as scientific research.

- **Appendix II** includes species not necessarily threatened with extinction, but which may become so if trade is not regulated. The trade of any part of the plant, or specified parts and derivatives are subject to controls, or specifically excluded. These species are listed with annotations.

Orchid herbarium specimens are subject to CITES regulations.

CITES Registered Scientific Institution label.

- **Appendix III** lists species that are threatened locally with extinction through trade and are subject to controls within certain countries.

Species, genera and families can be listed. The main plant groups listed include Orchidaceae, Cactaceae and Cycadaceae listed at a family level. The main plant genera listed are *Aloe*, *Aquilaria*, *Conophytum*, *Cyathea*, *Cyclamen*, *Dalbergia*, *Diospyros*, *Dipteryx*, *Encephalartos*, *Euphorbia*, *Galanthus*, *Gonystylus*, *Handroanthus*, *Nepenthes*, *Pachypodium*, *Pterocarpus*, *Rhodiola*, *Sarracenia* and *Tabebuia*.

Herbaria and the CITES Registered Scientific Institution Scheme

To facilitate research for science and conservation, only CITES Registered Scientific Institutions can exchange scientific material without having to apply for CITES permits. This applies to loans, donations and exchanges of specimens, including herbarium specimens, preserved, dried or embedded, and live plant material for scientific study. The CITES Registered Scientific Institution uses a label scheme to exchange material, at no cost to the institute.

Criteria for using the CITES Registered Scientific Institution scheme:

- Both institutions must be registered with their national CITES authorities
- Material must be legally acquired and accessioned into the institutions' collections
- Records are kept of loans and transfers

Checklist for CITES specimens

Determine whether the specimen is CITES-listed:

1. CITES species lists are available on the CITES website (www.CITES.org) and the UNEP-WCMC website Species+ (www.speciesplus.net).

2. Check the annotation for that species to determine whether the parts and derivatives are also subject to controls.

3. Use the label scheme for CITES Registered Scientific Institutions. Alternatively apply for CITES permits, check if your national CITES authorities offer fee waivers for scientific specimens.

4. Some countries have stricter domestic legislation: please check with the country of export and import before exchanging specimens.

5. If the species is not listed on CITES, it does not require a CITES permit but may need other documentation.

> **TIP**
> For more information, BGCI (2021) offers free CITES learning modules. https://www.bgci.org/resources/bgci-tools-and-resources/cites-learning-modules/.

FURTHER READING
CITES https://www.cites.org/
UNEP (2005, 2006, 2021).

BIOSECURITY
IMPORTING AND EXPORTING HERBARIUM COLLECTIONS
Alan Paton and Sara Redstone

Biosecurity is the application of knowledge and processes to reduce the risk of introducing non-native organisms to new areas as this can have significant negative impacts including damaged livelihoods and loss of ecosystem services and biodiversity. While herbarium specimens are normally considered no longer viable, particles on or in them may be, so any associated risks need to be managed appropriately.

Import
The movement of biological materials such as plants and plant parts, fungi and other biological specimens (e.g., plant pests and pathogens), soil and growing media, wood and wood products (including bark, logs, and packaging) and artefacts (including works of art, books, textiles, etc.) can carry significant biosecurity risks to herbaria and other collections such as libraries, illustrations or archives and can even pose risks to live plants and the wider environment.

Herbaria should have processes in place to safely manage the import, movement, use and sharing of biological materials – whether plants, DNA, herbarium specimens or artefacts. This is particularly important when dealing with invasive species, whether this is the plants themselves or related pests or diseases, which may include plant pests and pathogens such as nematodes, fungi, bacteria, viruses and viroids which have persistent resting stages that can survive drying and freezing.

Export
Different countries have different regulations surrounding the import and export of live and preserved plant and fungal material. Herbarium managers need to understand these regulations when importing and exporting collections into or from their country. Regulations for import may involve specific licence arrangements covering specific plant, pest or fungal species or plant parts (such as seed or fruits). These regulations are also likely to cover subsequent use of the material and how they are handled.

When considering export, managers should make enquiries to herbaria they

At Kew, herbarium plant material which is to be discarded goes into a large lockable skip before incineration.

intend to send material to, so that they know what measures must be taken to ensure the safe and secure transfer of material into that country. For example, some countries may require phytosanitary certificates or special import licences covering herbarium specimens. Whether for import or export, the relevant procedures and documentation should be understood and followed before sending material. It is likely that failure to follow procedures will result in the material being returned or even destroyed at the point of entry to the country.

There are often regulations covering use or transport of plant and associated material within a country. This may be focused on invasive species or plant species at risk of disease. Seed may be covered by more extensive regulations.

For example, in the UK, other than *Vitis* and *Solanum tuberosum*, which are prohibited, all seed imported requires a phytosanitary certificate from the country of origin. Seeds, spores or pathogen samples should NOT be removed from herbarium specimens for culturing, even if the herbarium specimens themselves are otherwise exempt from control. Import of herbarium specimens should not be used as a way of circumventing import controls on associated organisms or seed. Where there is a valid need to remove material for cultivation or culturing then permission must be sought from the herbarium manager and the relevant authority, and the material should be managed in a facility designed to safely and appropriately manage the potential risks that it poses.

Incinerator at RBG Kew.

REDUCING THE BIOSECURITY RISK

Alan Paton and Sara Redstone

Institutions should have processes in place to reduce the biosecurity risk. This section briefly lists some important practices to help guide managers in developing such processes. It is also important to consider the relevant national legislation and licensing regimes (Brown *et al.* 2018; Beeckman & Rüdelsheim 2020; Wondafrash *et al.* 2021).

During the collection of herbarium specimens

- Clean hands, footwear, tools and vehicles before setting off to collect specimens AND during collection, to prevent the unintentional introduction and spread of pathogens or invasive species

- Always work from areas of high conservation value to low conservation value, where possible, to minimise the risk of introduction or spread of plant pests, pathogens or invasive species to the most sensitive areas

- Avoid sampling material which is clearly diseased and may be infected with plant pests or pathogens

- Avoid collecting material that is in contact with soil wherever possible, cleaning off as much soil as possible on site if unavoidable

- If the intention is to collect plant pests, pathogens or invasive species there are likely to be further specific regulations or licence procedures, including risk assessments, which will need to be followed before the material can be safely and legally imported or exported

1 Washing footwear in the field.
2 Removing surplus soil from plant material when collecting herbarium material.

- Leave surplus plant material and any soil, bark or plant debris associated with the specimens at the site of collection
- Clean hands, footwear, tools and vehicles before leaving the collecting area if possible – leave soil and debris behind

On arrival of specimens at the herbarium

- Specimens are securely bagged and shelved in a "dirty area" and subsequently frozen
- Staff involved in study of unmounted or mounted specimens should ensure any plant debris generated by them is promptly disposed of using designated plant waste bins rather than adding material to general waste. Specimen debris or specimens being disposed of should be incinerated or otherwise disposed of in a biosecure manner (e.g., autoclaving)
- If material is covered by national regulations, it may need to be transported in a secure manner. For example, in the UK, licensed material needs to be transported in three layers of containment, one of which at least should be shatterproof

Staff responsibilities

- Anyone working with plant specimens must make sure their work area is clean and tidy and should clean up any plant debris as soon as possible, using the plant waste bins
- Anyone working with and using herbarium specimens should practise good hygiene and wash their hands with soap and water or use hand sanitiser regularly, especially if they work with high-risk taxa (including but not limited to Solanaceae, Vitaceae, Rosaceae, Umbelliferae, Fabaceae, Poaceae, Rutaceae, Fagaceae, Coniferales)

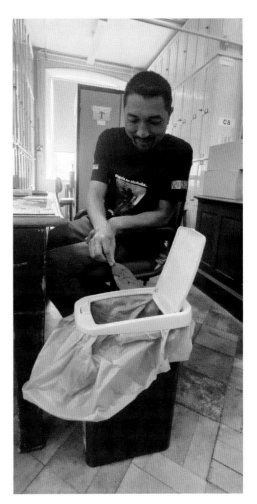
Specialised rubbish bin for disposal of plant material in the herbarium.

- Anyone handling herbarium specimens should avoid handling live plants immediately afterwards without thoroughly cleaning their hands with soap and water or alcohol-based hand sanitiser

EQUIPMENT
Zoë A. Goodwin

Equipment used for collecting plants can vary based on the types of plants you wish to collect and the ecosystem in which you are collecting. However, there are some fundamentals that all collectors need to consider.

Collecting the plant

To press plants, ideally you need a wooden press and two straps with old newspaper and cardboard or blotters. Traditional wooden presses can be expensive but should be very light and very strong.

To collect a specimen from tall vegetation you need pole pruners or a rope saw. Handheld secateurs are needed to cut the plant specimens to size and trim unwanted material. Permanent markers (if drying that night) or pencils (if collecting in alcohol) are needed to label the newspaper. Additionally, jeweller's tags labelled in pencil can be used to label the specimen directly.

Think about additional material you might want to collect and what equipment that would require: sealable tubs, teabags, and silica gel to collect samples for DNA extraction, airtight bottles, and alcohol to collect flowers in spirit, paper packets for fruit, bryophytes or lichens, waxed paper for fungi or delicate flowers.

Recording data

Use a waterproof notebook and a pencil or waterproof pen to record your data.

> ### TIP
> Tie a short piece of brightly coloured flagging tape to equipment like the secateurs and pens; this makes it much easier to find this equipment when you have you all of your kit laid out on the ground.

If you are recording your data in an electronic device, bring a waterproof notebook and a pencil as a backup.

Think about the data you want to record and what you will need to record it – GPS, clinometer, tape measure and digital camera. Electronic equipment needs spare batteries, memory cards and waterproof bags (dry bags or sealable ziplock bags).

If you have been collecting before, remember to note down the last collector number that you used.

What to wear and keeping safe

Plan your collecting. Consider exposure (cold, heat or the sun) and insects when deciding what to wear. Wear long sleeves, long trousers and consider a head covering, for example a hat with a wide brim in sunny locations, a woollen hat in cold locations or a small hat or bandana around the head in a forest. Wear shoes appropriate to the environment. Do you want to keep your feet dry? Will you need good ankle support for slopes and rocks?

Think about additional items to keep you safe, such as sunscreen, insect repellent, first aid kit, map, spare cash and a mobile phone or satellite phone. Finally, bring water to drink and emergency snacks and rehydration salts.

FURTHER READING
Liesner (2017).
http://www.mobot.org/MOBOT/molib/fieldtechbook/welcome.shtml.

A large amount of equipment is required for a day's plant collecting. The exact amount of equipment will vary depending on the additional material and data you want to collect and the environment in which you are collecting.

FIELD NOTES
DIGITAL COLLECTION
Justin Moat and Steven Bachman

When collecting, it is important to record data to accompany the physical specimen(s) – this gives context to the specimen and supports a wide range of research (James *et al.* 2018). It will allow those using the specimens in the future a greater understanding of the species in the field, including its ecology and habitat.

Recording field data electronically
With the progression of technology specimen and field data can be recorded either digitally using electronic devices (including smartphones), physically on paper, or even as a hybrid of paper and digital. Each of these methods have their own advantages and disadvantages, some of which need to be managed prior to, during and after field work.

There are numerous electronic devices and applications which allow the recording of field data in a digital format, which can be imported into a database, spreadsheet etc. These can range from simple note-taking applications on a smartphone to a dedicated application for field data collection.

We recommend the Open Data Kit (ODK), which is a highly flexible open-source set of software for collecting data in the field (https://opendatakit. org/software/). It allows the collection of images, sketches, GPS, audio and video as well as textural data, and can be deployed on smartphones, tablets and laptops. The advantages of a 'born digital' (i.e., data entered directly into databases, spreadsheets etc.) specimen workflow are gains in efficiency and consistency in data entry (using controlled vocabularies) and avoidance of transcription errors through direct transfer of data to collection management systems (Powell *et al.* 2019).

Electronic field data collection on android device, using Open Data Kit (ODK).

Recording field data, electronically – considerations

Collecting digital field data is often preferable, but it does come with limitations which need to be planned for, with the following particularly important:

1. **Waterproofing**. Equipment needs to be either waterproof or stored in a waterproof pouch (note, wet electronics can be dried with silica gel).

2. **Battery life**. Even the best devices will eventually lose battery power. Where electricity may be limited, make plans to recharge devices wherever electricity is available and utilise any battery-saving options on your devices such as airplane mode to turn off wi-fi antenna and reduce screen brightness. Bring extra batteries and/or portable power banks. Finally, explore the use of alternative sources of energy such as solar chargers.

3. **Failure**. Any electronic equipment can fail, sometime with dire consequences, and we recommend the following contingencies are put in place:

a. Backup units are made available and should be spread within the team in case one bag is lost or stolen.

b. Data are backed up daily. With internet connections it is easy to back up to the cloud, but in remote areas backups should be made every day and on at least two devices (either memory cards or hard disks), and these three backups are stored with different individuals. Leave one backup in country as a safeguard. If all else fails always have a notebook to hand as well (see overleaf).

With the above in mind is would be prudent to purchase high-quality electronic devices with large capacity batteries, along with multiple memory cards – preferably all waterproof.

Botanist capturing digital notes and high-resolution images of a plant in the field in Madagascar.

PHYSICAL COLLECTION: NOTEBOOKS

Justin Moat and Steven Bachman

Paper notebooks are simple, cheap and do not suffer from failure or loss of power. They lack the functionality of an electronic device and can suffer from handwriting and transcribing issues. Additionally, if digital data are needed, they will need to be transcribed again at some point, which can introduce errors.

Recording field data, notebooks – considerations

1. Collecting books should be waterproof, sturdy, and preferably bound, providing a firm writing surface. Pocket-sized is most convenient. Many herbaria design and produce notebooks to their own specification.

2. Pencils and pens. HB-2B are the best; brightly coloured ones are less likely to be lost. Propelling pencils are convenient as they will not need sharpening but take spare leads. Ballpoint pens may also be useful but do not use pens with ink that is not water-resistant.

3. Backup. The easy way to "back-up" notebooks is to photograph each page. As with electronic devices this should be done at least daily. Additionally, record your name and address at the beginning of the notebook in case it is lost.

What data to record

Data should be recorded as completely and methodically as possible and preferably at the time of collection. Write carefully and clearly in pencil if using notebooks and do not use obscure abbreviations. The bare minimum to be collected are:

- Locality (GPS reading and place name)
- Habitat
- Description of plant
- Collector's name and number
- Field identification (even if only to family or group)

The full data recommend and details to be recorded are as below; additional data can be recorded depending on the purpose of field work or study.

Kew's herbarium collection notebook. **1** Front page. **2** Reminder of what to record in the field. **3** Location and specimen data. **4** Description and additional notes.

1

2

REMINDER TO COLLECTORS

PLEASE:

Collect complete, representative specimens in flower and/or fruit and include underground parts where possible. Put specimens in press at time of collection and dry quickly or use alcohol method.

Indicate locality precisely in relation to some easily located geographical feature (e.g. large town) and give co-ordinates as degrees/minutes/seconds, noting GPS datum (e.g. WGS 84) and accuracy.

Describe habitat and vegetation carefully and as fully as possible.

Estimate the abundance of the species at the collection site – give numbers of individuals per unit area or use the ACFOR scale (**A**bundant, **C**ommon, **F**requent, **O**ccasional, **R**are). Document the observable threats to the species at the site (e.g. harvesting, agriculture, grazing, fire, development). Where possible, describe the intensity of the threat and the demographic structure of the species at the site (e.g. few reproductive adults seen, no regeneration observed).

Make full descriptive notes **at the time of collection**, emphasising features liable to be lost or obscured in the collecting and drying process, e.g. habit, size of plant, dbh, exudate, flower colour/scent.

Number collections in a single consecutive series. Cross refer to ancillary collections, e.g. images, alcohol material, etc.

Write carefully and clearly **in pencil** and do not use obscure abbreviations.

3

LOCATION DATA

Country (province, district, etc.):

Locality:

GPS datum/accuracy: Altitude (m):

Lat.: N / S   °   '   " Long.: E / W   °   '   "

Habitat/vegetation:

Geology/soil:

SPECIMEN DATA

Field det.:

Vern. name:

Abundance:

Threats (type/level):

Uses:

Collector:

Number: Date: /   /

Collected with:

Images:

Material: Herb.   Carp.   DNA   Wood   Seed   Alc.   Live   Other:

Duplicates:

4

Description (habit, height, leaves, flowers, fruit etc.):

RECORDING SPECIMEN DATA

Justin Moat and Steven Bachman

When collecting, it is imperative to accurately record relevant data: primarily these are i) locality information, ii) specimen data, and iii) plant description. While specimens are usually collected for taxonomic and systematic research, all these data are useful for studies such as extinction risk assessment, phenology and habitat characterisation.

Typical equipment for field notes. **1** Portable power bank. **2** Mobile phone for data collection. **3** GPS unit. **4** Field notebook and pencil. **5** Tape measure and/or ruler.

Locality data

Locality data provide verifiable and citable evidence of the occurrence of particular plants at particular points in space and time. These are very important when using geo-referenced points for estimating conservation metrics, such as the Extent of Occurrence (see IUCN 2012).

Location information

Indicate locality precisely in relation to some easily located geographical feature (e.g., large town) and give GPS coordinates as degrees/minutes/seconds, noting GPS datum (e.g., standard is WGS 84), altitude and accuracy (in metres). If a map or another method is used to determine latitude and longitude, this should be recorded (i.e., location from Map 1:250,000).

Tips for getting more accurate locations from GPS units

- Units should be held high so that the signal is not shielded by the body
- Units should be given time to receive a good signal (especially if you have moved some distance since last reading)
- Do not turn off the GPS between collecting sites
- Care should be taken to set correct formats of coordinates; many GPS units and applications default to decimal degree or decimal minutes, which should be avoided as can be difficult to transcribe
- Dedicated GPS units can record individual waypoints. It is recommended to use these for each collection and record the waypoint ID on your electronic or field notebook. It is also useful to record tracklogs, which can be an aide-memoire, for example to retrace your route, and can also be used to geo-reference images from digital cameras

Habitat/vegetation

Describe habitat and vegetation carefully and as fully as possible; use national or accepted standards.

Relative abundance

Estimate the species abundance – give numbers of individuals per unit area or use the ACFOR scale (Abundant, Common, Frequent, Occasional, Rare).

Threats

Document the observable threats, and their intensity, to the species at the site (e.g., agriculture, grazing, fire, development). If possible, record the demographic structure of the species (e.g., few reproductive adults seen, no regeneration observed).

Specimen data

Specimen data can be considered as the unique identifiers for each specimen, linking the specimen number to other data via a collecting number, collecting data and the collector's name; together these provide unambiguous combination that can be used when citing specimen collections in publications and databases.

Field identification

Family, genus, species if known, even if not to group (i.e., fern); this is useful when processing specimens after an expedition to ensure numbers and dates do not get mixed against different plants (if, for example, several collectors are making collections on the same trip).

Vernacular name

Record name phonetically if the correct spelling is uncertain and record dialect used; try to check reliability of name, and if using electronic device record the spoken name (with permission).

Uses

Record any uses of plant; obtain permission from relevant stakeholders and try to get confirmation from a local collaborator [within the confines of Access and Benefit Sharing legislation].

Collector/s

Name of collector (or collecting team).

Collector number

Unique collection number – easiest to start with a running sequence to ensure the name + number combination is not repeated. Some collectors use combinations of years and numbers as a composite collector number.

Collection date

Date of collection: day/month/year.

Images

File names or ID; also if needed, identify whose camera was used.

Material

Type of material collected; Herbarium specimen, bulky collection, DNA sample, Wood sample, Seeds, Alcohol collection, living collection, other.

Duplicates

Herbarium codes for duplicate material (if these are known; often these are updated collections that have been processed rather than at the time of collection).

Plant description

Herbarium specimens are by their nature a flattened and dried-out sample of a plant that would have been fleshy with a 3-D structure, with the leaves and flowers colourful *in vivo*. A description of the key attributes that are lost on pressing and drying is invaluable for identification and taxonomy once the specimens are in the herbarium. Some of the key attributes to be considered for recording at the time of collection are listed below, but this is not definitive or proscriptive as different plant groups have different key characters to be noted.

Habit

Tree (including overall shape; see Whitmore 1972), shrub, climber, epiphyte (and type of epiphytic habit), herb etc.

Underground parts (especially important in monocots)

- Tap root, fibrous roots, tubers on roots, extent of roots
- Scent of cut parts
- Rhizome – depth in soil, length, spacing of shoots
- Bulb, corm or tube – size and shape

Stems and trunks

- Size: total height, girth at breast height (GBH, usually in m) or diameter at breast height (DBH, usually in cm); remember to note which metric was used (GBH or DBH), and if it was estimated or directly measured
- Height of trunk (m) or stem before branches; whether buttressed or not
- Bark colour, texture, thickness, lenticel colour
- Wood hardness, colour, grain type
- Diameter at various heights (m) (if variable)
- Cut trunk ('slash'), sap or latex including colour, smell, consistency, and other properties
- Shape in cross-section (circular, fluted etc.)
- Internode lengths
- Thorns, spines, especially if on trunk (as usually only branch ends are collected)

Leaves

- Deciduous or evergreen
- Texture, colour(s), smell, glossiness, colours on upper and lower surface (e.g., glaucous in Lauraceae)
- Exudate or glands
- Orientation in relation to petiole or stem etc., e.g., 'pendulous', 'horizontal'
- Large and/or compound (collect sequentially and label each part)
- Outline shape (if large or complex), sketch
- Note if heterophyllous (juvenile, shade or submerged aquatic leaves)

Inflorescence

- Exudate or glands
- Cauliflorous, ramiflorous, any other data on position or form that may be lost in prepared specimen
- Colour of axis

Flowers

- Note if heterostylous or if plant monoecious or dioecious (do not mix collections if plants unisexual!)
- Scent
- Corolla colour, texture
- Calyx colour, texture
- Exudate or glands
- Behaviour (e.g., open early and closed by 12 noon)
- Pollinators and/or floral visitors

Fruit and seeds

- Smell
- Colour, texture
- Size, shape, especially of fleshy/wet/ bulky fruits
- Seed-coat colour, texture
- Aril colour, texture
- Dispersal (animals, wind or water)

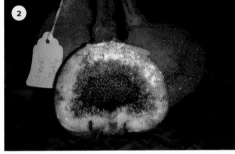

Good field notes are required for all plant characters that will be lost on drying, such as exudate and flowers and fruits when not associated with leaves. **1** *Tinomiscium petiolare* with free-flowing white exudate from the stem. **2** *Ficus* sp. with sticky exudate from the 'fruit' wall. **3** *Macrosolen formosus*, ramiflorous with numerous flowers along the stem. **4** *Epicharis cuneata* (*Dysoxlum cauliflorum*), cauliflorous with mature fruits on the trunk.

COLLECTING TECHNIQUES

Marie Briggs and Kipiro Damas

Herbarium specimens are a key data source for research in disciplines such as conservation, climate change, ecology, ethnobotany, molecular systematics and plant taxonomy. To maximise their utility, they must be correctly identified. Accurate identifications are more likely in 'fertile' specimens with informative labels.

There are various ways to make good quality herbarium specimens. The method chosen will differ according to individual and institutional preferences, environmental conditions, equipment available and proximity to facilities such as electricity. Here we will discuss some of the methods commonly used.

1a–c Collecting process.
2–4 Collection methods: trowel, long-pole pruner, tree climbing.
5 Specimens should be fertile.
6 Underground parts attached.
7 Young and mature fruit.
8–9 Using a field press and bag before pressing.

Herbarium specimens – what to collect

- Carefully observe: are the leaves compound/dimorphous; do they or the flowers/fruits differ dramatically in size; is the plant monoecious, what is typical for the population etc., and represent this in the selected material
- Specimens should be 'fertile' (with flowers and/or fruit) and should include leaves (if present) and stem and/or branch material
- 'Sterile' material (with no flowers or fruit) and solitary leaves, half leaves and fragments are generally undesirable
- If practical, include underground parts (roots/bulb/corms, etc.). These can be sliced into layers if they are too thick

Herbarium specimens – how to collect

- Each specimen should be given its own unique collection number [see **TIP**]
- Aim to collect 3–5 duplicates per specimen (see: Collecting Duplicates):
  - For small plants, a few individuals of the same taxon from the immediate vicinity can be included together to form one specimen. Care should be taken to avoid over-collecting
  - For larger plants, duplicates should be from the same individual. Specimens collected from different individuals or from the same individual at different times should be given separate collection numbers

- Use a trowel to carefully extract smaller plants from their substrate with underground parts attached. Remove excess soil
- Use secateurs or similar equipment to cut representative samples of larger plants, preserving branching patterns as much as possible
- Long pole pruners, tree climbing experts and slingshots, amongst other methods, may be employed to collect out-of-reach plant material
- Place each specimen on one side of an open, folded sheet of paper – newspapers are often used as they are absorbent, tend to be a similar size to the drying press/resultant herbarium sheet and are generally widely available, but other absorbent paper may be use
- Some collectors prefer to place collections in plastic bags, laying material out in newspapers later. **PROS:** quicker, can collect more material. **CONS:** not suitable for specimens with delicate plant parts, specimen parts may get mixed up
- Some collectors attach jeweller's tags to specimens. If using: Write the lead collector's initials followed by the unique collection number on the tag clearly in pencil ('HB' to '2B' hardness works well)
- Avoid attaching tags to the end of stems where they can easily slide off

---

TIP

Every specimen should be given a unique collecting number (note duplicates, as detailed above, will share the same collection name and number). The simplest method is to start at number one and continue in a running sequence. The number and collector's surname become a unique identifier for that collection and, as such, the same number should never be used twice by an individual collector.

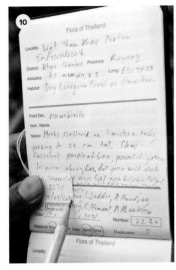

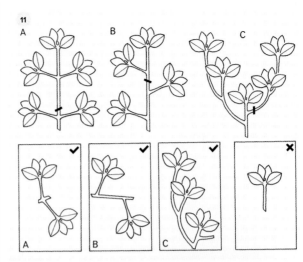

10 Notebook.
11 Preserve branching patterns. Specimens can be:
12–13 Folded.
14 Trimmed.
15 Split over several sheets.

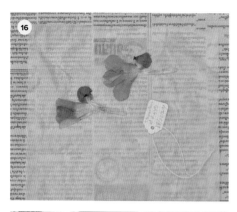

16–17 Dissections of flower and fruit (press delicate flowers in non-stick paper). **18–20** Field presses can be purpose-built or improvised.

Preparing plant samples for pressing and drying

- Spread plant parts out as best as possible to avoid multiple layers (which may slow the drying process and allow mould to form)
- Cut, trim or fold specimens to fit the dimensions of the intended herbarium sheet (sheet size varies between institution but is generally 26–30 x 41–45 cm)
- When removing leaves/leaflets, keep the petioles/petiolules and lamina bases, for information
- Specimens too large for one sheet may be split over several sheets Clearly indicate the sheet number on jeweller's tags or on the paper (e.g., sheet 1 of 3, sheet 2 of 3, sheet 3 of 3), under the same collector's number
- Display different aspects of flowers and fruits where possible, to maximise the information available on each specimen
- Make sure upper and lower leaf surfaces are showing – turn some leaves over or fold part of a leaf over.
- Include extra sectioned flowers and fruits where possible
- Make specimen as flat as possible, keeping any bulky parts such as large fruits and thick wood samples separate, tagged with the collection number. These can be dried separately (see: Collecting bulky specimens)

- Alternatively slice thicker parts in half or in sections for inclusion in the drying press
- At this point, the newspapers can be closed and put directly into a drying press (see: How to build a press), placed into a field press for later transfer to the drying press, or bundled up ready for the application of alcohol (see: Schweinfurth method)

TIPS

- Delicate flowers can be protected by pressing them between various types of thin paper – toilet tissue or greaseproof paper for example (see Fig. 16)
- 'Field presses' can be anything from specialist nylon collecting satchels and wooden herbarium presses to simpler cardboards and rectangles of plywood held together by straps or rope; See Fig. 18–20)

FURTHER READING
Womersley (1981); Victor *et al.* (2004); RBGE (2017).

THE COLLECTION OF SPECIAL PLANT GROUPS
ALGAE AND BRYOPHYTES
Joanna Wilbraham

The following plant groups require particular attention and techniques to make scientifically valuable specimens.

Algae

A good low tide will reveal much of the shore's seaweed inhabitants. Deeper subtidal habitats will require divers to sample. Collecting should be focused on living seaweeds attached to a substrate. If unattached material is collected, then it should be annotated as drift. Specimens can be collected directly into labelled plastic bags or vials. Charophytes can be collected directly by hand in shallow waters. A grapnel can be used for sampling in deeper water, but care must be taken not to damage underwater habitats unnecessarily.

It is possible to keep seaweeds and charophytes fresh for a short while by storing in plastic bags (with any excess water drained off) in a fridge.

How to preserve these specimens in a naturalistic form:

- Float the specimens out onto a sheet of drying paper before being placed in the plant press. Seawater is needed for more delicate seaweeds whose cells may burst in freshwater
- Annotate the paper sheet with the collection's details in pencil, and place it in the tray
- Float the specimen above the paper. Its features can then be spread out and arranged
- Carefully slide the paper with the specimen laid on top out of the tray, and place it in a plant press on a layer of absorbent papers
- Cover specimens with a sheet of nylon, muslin, or cotton fabric to prevent them sticking to the absorbent papers above
- Very lime-encrusted charophytes should be dried with very little pressure to avoid fragmenting the specimen
- To avoid the risks of the seaweeds starting to decay in the press it is essential to replace the absorbent paper with dry sheets approximately 12 hours after first pressing the specimens and on subsequent days until the specimens have dried. If conditions for drying are poor, then it will be necessary to expose the press to a gentle heat source, though be aware that excess heat can damage the specimens

FURTHER READING
Dipper (2016); John *et al.* (2011).

Bryophytes

The bryophytes include the mosses, liverworts and hornworts. Care should be taken when sampling bryophytes so as little disturbance as possible is made to the population; for example, by taking a few shoots from the edge of a mossy cushion, never the centre. Specimens are best collected directly into paper packets on which collection details can be written. Specimens should be left to dry with the collecting packets partly opened in a well-ventilated place. Artificial heat should only be used if necessary and always at a low temperature. Leaves supporting epiphytic liverworts should be dried under light pressure (as for vascular plants).

FURTHER READING
Paton (1999); Smith & Smith (2004).

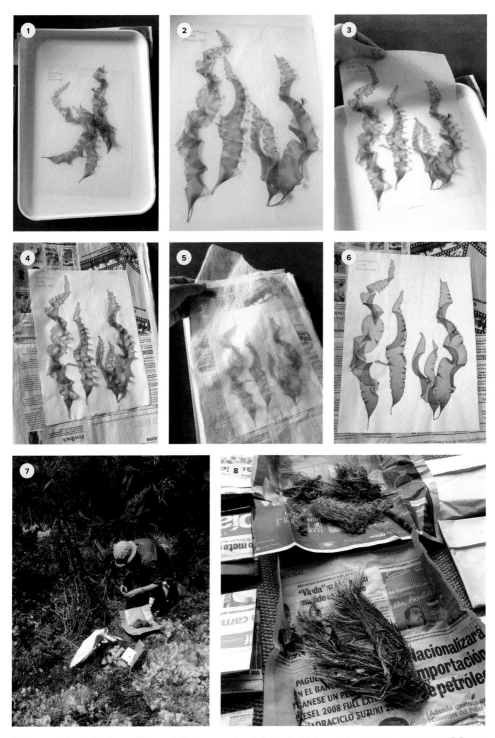

1 Seaweed floated in tray with annotated paper sheet. **2** Floated into a naturalistic arrangement. **3** Paper with seaweed carefully slide from tray; may help to angle the tray to slide the paper out of the water in the shallow end. **4** Place onto absorbent papers in press. **5** Covered with a sheet of porous non-stick material. **6** The dried specimen. **7** Collecting bryophytes. **8** Drying bryophyte specimens.

PTERIDOPHYTES AND PALMAE
Anna Haigh, John Dransfield and William J. Baker

Pteridophytes

Ensure ferns and fern allies are fertile before collecting; ideally the spores should be ripe. Spores shed during the drying process should be kept in a packet. Some groups also reproduce vegetatively with buds on the lamina or rachis; these should be preserved if present.

Delicate rhizomes should be carefully pulled from the substrate such as moss or bark to ensure the specimen is not dried in a mass that cannot be untangled once dry.

Medium-sized ferns can be treated like most other plants. However, take care to keep the base of the petioles and a portion of rhizome, as the scales and hairs found here are important taxonomic characters. If the rhizome is creeping, enough should be taken to indicate how the fronds are spaced.

It is impractical to collect entire fronds of tree ferns and other large ferns. It is sufficient to collect the following:

- Petiole from base to first pair of pinnae
- Central portion of lamina, including the rachis
- Lamina apex
- Duplicates must be from a single frond of the same plant
- If it is a tree fern, and is common, cross sections of the trunk are useful. Alternatively, a silicone moulding technique can be used to preserve trunk surface structures without damaging the plant
- A young unfolding leaf can be dried or preserved in alcohol

FURTHER READING
Janssen (2006).

Palmae

In most cases, there is little point in collecting a palm that has neither flowers nor fruit, with the exception of rattans, which can make informative sterile specimens.

A small understorey palm can sometimes fit on an herbarium sheet. For the majority of palms, however, a specimen will consist of numerous items sampled from representative parts of each organ. Take care to label every item with a tag.

- Collect a sample of stem with at least two nodes represented. A thin strip of outer stem will suffice for robust palms
- Split the leaf sheath longitudinally, folding the material, or preserving the upper and lower portions if too large
- If the leaf is small, collect it whole, folding it to fit an herbarium sheet. In most cases, however, you will need to cut it up
- For pinnate leaves, take a portion of the base of the petiole, a portion of the top of the petiole with the first leaflets, a middle section with rachis and leaflets, and the tip of the leaf. Remove the leaflets from one side of the rachis if they are too big. Fold the remaining leaflets in a manner that allows them to be easily unfolded for study
- For fan leaves, take a portion of the base of the petiole, a portion of the top of the petiole, including the hastula and lower segments on one side, then remove a portion from both the centre and the side of the leaf. If the inflorescence is too large to collect whole, cut it to preserve a basal portion, a mid-section and a tip. Try to retain evidence of the maximum number of branching orders and preserve bracts. Make

sure you have good flowers and/or fruit. You may not be able to obtain both from a single palm

- If flowers are not visible, you may be able to find some inside a closed inflorescence bud. If no ripe fruits are available, you may be able to find fruits seeds or germinating seedlings on the ground
- Spirit material of palm flowers and fruit can be invaluable
- Rattan collecting is similar, but it is important to collect the climbing whips and to note whether they arise from the leaf tip (cirrus) or

leaf sheath (flagellum). Do not remove the leaf sheath from the stem or split it in any way. Simply, cut a length of stem with sheaths, including at least one leaf base

FURTHER READING
Dransfield (1986); Baker & Dransfield (2006).

TIP
Take meticulous notes so characteristics lost in the collecting process (e.g., leaflet number, stem height, leaf length) can be interpreted in the herbarium.

1 A tree fern collection should include the base, a central portion and the apex of a frond. **2** Ensure ferns are fertile before collecting. **3** Carefully label each part of the fern specimen, to keep a whole collection together. **4** Detailed notes of palms and pteridophytes are essential to record details that are lost in the collection process. **5–6** Parts of a palm specimen prior to drying. **7** Parts of palm inflorescences preserved in spirit.

ORCHIDACEAE AND CACTACEAE
Anna Haigh and Daniela C. Zappi

Orchidaceae

Thick pseudobulbs and tubers of terrestrial orchids may need to be cut longitudinally to aid drying and prevent the specimen from being too bulky. Succulent or fleshy leaves can be cut superficially on the underside to aid drying if necessary. It is important to make spirit collections of the flowers. Fragile flowers must be pressed rapidly to prevent wilting before putting in the press, ideally using tamping paper to prevent fleshy tissues adhering to more traditionally used absorbent papers. Colour photographs particularly of the flowers, but also of the specimen *in situ* are desirable. It is often useful to photograph the lip and the ventral side of the column separately as these may have important characters that are otherwise impossible to see in a photo. Some orchids (e.g., *Nervilia*) do not have flowers and leaves in the same season. In this case it is desirable to make sterile specimens as well as fertile (but different gatherings must be numbered separately).

1 Photographing orchid flowers is important to show colour and structure. **2** Orchid flowers are best preserved in spirit to keep their structure. **3** Collecting Cactaceae in the field. **4** Cutting Cactaceae stems in preparation for pressing and drying. **5** Pressing Cactaceae flowers. **6** Make both transversal and longitudinal sections to give an idea of shape and number of ribs and areoles, as well as aspects of the epidermis.

FURTHER READING
Taylor (1991).

TIP
Once cut, Cactaceae tend to go mouldy within about a week if not dried. Aim to get specimens both dry and flat while preserving their morphology as much as possible.

Cactaceae

The best way to preserve the structure of flowers, fruit and whole stems of small plants is in spirit. Drying Cactaceae needs a more creative approach, and they can be prepared by a variety of methods.

Key features to preserve are:

- No. of ribs
- Stem diameter
- No. and positions of spines in each spine cluster
- Size and shape of areoles
- Flowers (both inner and outer parts)
- Fruit and seeds
- If the cactus is small, collect the whole plant
- If it is a large cereoid cactus, aim for a fertile branch

How to collect Cactaceae specimens:

Tongs can be very useful, with cardboard or even a piece of carpet to protect your hands. If the plant can be transported back to lab conditions in a live state, it can be kept at least two weeks in dry paper (prepare flowers or fruit separately as these will not keep). Once in a position to take time, the plant will need to be appropriately sectioned to show the important features. It is very important to make both transversal and longitudinal sections to give an idea of shape and number of ribs (transversal) and shape of ribs and areoles, as well as aspects of the epidermis (longitudinal section). The parenchyma can be removed from longitudinal sections of very fleshy specimens or, if your cactus is thinner (like some epiphytes), it can be scored with a knife to aid drying. Smaller parts such as flowers and fruit can be dried separately.

Use plenty of corrugates, cardboard, and pieces of plywood to keep them flat and prevent damage to adjacent specimens. In a good dryer, with regular changing of damp dryer materials, the specimens should dry quickly.

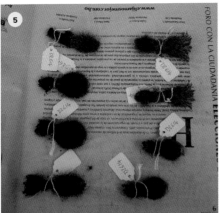

HOW TO BUILD A PRESS
Nina Davies

Once a plant collection has been made, the specimens are then placed into a press for preservation. A wooden press provides strength and equal pressure across the press through the drying process. Here are the instructions for preparing and building a wooden press including how to layer specimens.

STEP 1

Gather your equipment and prepare specimens according to Collecting techniques. Press with lattice, straps, corrugates, blotters, dry newspapers/absorbent paper.

STEP 2

Place one side of the press down flat with the long slats facing downwards and short slats facing upwards (this will be on the inside of the press). This helps to distribute the pressure evenly over the press when it is strapped up and protects the lattices from damage.

STEP 3

Place a pair of open straps underneath this side of the press as this will make it easier to strap up the bundle in step 12. Placing the clasps in alternating directions will help to tighten the press evenly.

STEP 4

Place one corrugate in the press. The corrugates can be aluminium or cardboard, but the direction of the corrugation should be parallel to the short end of the press to enable air flow through the press.

STEP 5

Place a blotter on top in order to absorb the moisture of the pressed plant.

STEP 6

Place an open double sheet of newspaper or other folded absorbent paper on top of the blotter. Plant material can now be added.

STEP 7

Place one collection in the newspaper (which may be one or multiple individuals). Turn leaves and flowers/fruits over to show various aspects and hold in position as the paper is closed.

STEP 8

Quickly add a blotter on top of the closed newspaper.

STEP 9

Continue in the same way through steps 4–8, creating a sandwich effect with the plant material inside: corrugate, blotter, specimen in newspaper, blotter. Make sure the folded side of each newspaper is on the same side so no plant material falls down onto the heat source. The folded side should always be closest to the heat source.

STEP 10

Finish your press with one corrugate.

STEP 11

Place the other half of the press on the top with the short slats facing downwards and the long slats facing upwards.

STEP 12

Pull up and fasten the straps while holding the press steady. Apply strong downwards pressure to the bundle while tightening the straps as much as possible, flattening the specimens in the process.

As the specimens dry, check the press at intervals and change any wet newspapers and blotters; this is a good time to rearrange the specimens if needed.

A fabric field press can be used as a temporary measure in the field when collecting and travelling until the specimens can be transferred to a traditional press. This folds around the bundle but be sure to use cardboard corrugates to not damage the fabric.

TIPS
- Keep in mind the rough size of the herbarium sheet when placing plant material in the press. Herbaria have different sized sheets
- If there are not enough corrugates available, space these evenly throughout the press

STEP 1

STEPS 2–4

STEPS 5–6

STEP 7

STEP 8

STEP 9

STEP 10

STEP 11

STEP 12

FABRIC FIELD PRESS

METHODS FOR DRYING PLANT SPECIMENS
Marie Briggs and Kipiro Damas

Specimens collected from living material may be preserved indefinitely by pressing and drying, and storing in a dry, pest-controlled environment. The methods used will depend upon environmental conditions and personal preference. A selection of methods (by no means exhaustive) is detailed below.

Stoves – gas and fuel
Various types of portable stoves can be used to dry plant specimens, including gas and kerosene. Use a frame or pulley system to suspend the presses c. 1 m above the flames. Ensure the press straps are secured away and add a barrier, e.g., chicken wire or a metal plate, to prevent material falling onto the flames. When the stove is lit, wrap a thermal blanket or tarpaulin around the outside of the frame, securing it away from the flames. This will direct heat up through the presses, drying the specimens in the process. Aim for a medium heat, not exceeding 50–60°C. For reference, one 'Campingaz'-style, 190 g non-reusable tank can burn up to 12 hours if the flame is kept at the lowest setting (X. van der Burgt, *pers. comm.*).

PROS
Most specimens will be dry within 24 hours, freeing the drying presses for the next batch of specimens; both DNA and the natural colour of the specimens tend to be well preserved; suitable for use in humid environments.

CONS
Reliant on having enough fuel, which can be tricky on remote collecting trips; using too high a heat can reduce the quality of extractable DNA and burn specimens or make them overly brittle; fire can be dangerous!

Sun
In dry environments, specimens can be dried through frequent changing of newspapers and blotters. Place presses in a warm, dry spot. Check and frequently change papers and blotters until specimens are dry (see TIPS). Damp paper coming out of the press can be dried in the sun, for re-use (J.R.I. Wood, *pers. comm.*).

PROS
Papers and presses are the only equipment required for drying; both DNA and the natural colour of the specimens are usually well preserved.

CONS
Mould can form quickly if papers are not changed frequently enough; large amounts of newspaper and presses may be required, depending on the amount collected and speed at which specimens can be dried; not suitable for humid environments.

> **TIPS**
> - Presses should be checked every 12–24 hours (at least 6–12 hours, for the 'sun' method, and more frequently when drying fleshy plants or in extreme heat to avoid mould forming, swapping damp blotters and newspapers for dry ones)
> - If there are not enough corrugates, space these through the press

1–2 Press and bulky fruit drying in sun.
3–4 Metal drying frame, kerosene stoves.
5–6 Presses suspended by pulley system.
7 Wire barrier and gas stoves.

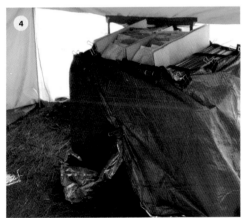

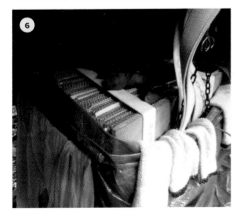

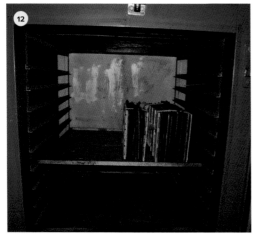

Electric fan heater

Some plant collectors have adapted standard household electric fan heaters to dry plant specimens. Place the press c. 1–2 m away, with the long edge towards the front of the fan heater. Create a 'tunnel' between the press and the heat source with polythene, a tarpaulin or something similar, and seal it so the air flows down the tunnel and is forced out over the press. For heaters of c. 1000 watts, set the heat to half power (adjust accordingly for other wattages), and turn the thermostat to 'full'. It is important to use safety-tested extension cables and equipment for this and to hang a smoke detector nearby, as a safety precaution. Regularly check the plug to ensure it is not heating up (X. van der Burgt, *pers. comm.*).

PROS

As per gas and fuel methods.

CONS

Need to be close to a reliable electricity source, potentially high energy consumption (payment for extra electricity usage should be factored into budget), potential fire hazard!

Post drying

Once dry, check to make sure jeweller's tags are present and correct, then write the collector's initials and number on the outside of all newspapers in marker pen. Bring all duplicates together. These simple steps will make sorting material and adding the corresponding labels easier.

Pack completely dry specimens in bundles, between two sheets of cardboard and tied with string using an herbarium knot. Place in a strong waterproof plastic bags and seal with tape. Dry specimens are brittle so pack in strong boxes if they are to be transported, to save them from damage.

Oven drying

If specialist equipment is available, specimens may be dried in drying ovens. These come in various shapes and sizes but are essentially large electric ovens. The temperature should be set at a steady 40–60°C, and presses placed on their sides, on a shelf to dry.

8–10 Electric fan heater.
11–13 Electric oven.

THE ALCOHOL/'SCHWEINFURTH' METHOD FOR PRESERVING PLANT SPECIMENS

Marie Briggs and Kipiro Damas

If specimens cannot be adequately preserved through drying in the days after collection, it is possible to suspend the need for drying for several months using the alcohol, or 'Schweinfurth' method (modified from Schrenk 1888).

Temporary preservation of specimens using alcohol

- Prepare specimens (see: Preserving specimens)

- Newspapers containing specimens should be piled up on top of each other to a maximum height of c. 25 cm

- Gently compress the bundle flat and wrap an extra layer of newspaper around the outside (this helps to keep loose parts from falling out)

- Tie the bundle firmly and place in a strong polythene bag with no holes

- Working in a well-ventilated space, pour c. 500 ml of c. 70% alcohol on the specimens, ensuring the specimens and papers are uniformly damp

- Seal the plastic bag securely with strong tape (e.g., parcel tape) or string

- Any liquid in the bottom of the bag can be redistributed to the specimens by gently turning the sealed plastic bag

- Bundles can be effectively stored for months, but the integrity of the plastic bag and seals should be regularly checked

Considerations

- This method is particularly useful for remote work in wet and humid environments

- Alcohol percentage is important; too low and the specimens will quickly become mouldy, too high and the specimens will be brittle and may lose their leaves during the drying process

- As alcohol can affect specimen colour (often turning them brownish) and can make later extraction of DNA more difficult (Forrest *et al.* 2019), make good field-notes and collect samples in silica gel for DNA studies where possible

> **TIP**
> Pad spiny specimens with extra newspaper and put them in the middle of the bundle, if possible, to prevent piercing of the plastic bag.

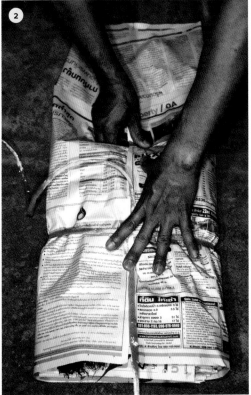

1–2 Prepare and wrap specimens in bundles.
3 Place in strong plastic bags and add alcohol.
4 Seal securely.

THE COLLECTION OF DUPLICATES

Laura Pearce

Duplicates are additional specimens made from the same plant collected at the same time by the same person. They are made as a form of insurance in case the first duplicate is lost or damaged and to ensure that a collection is available to a wide group of researchers. Each duplicate is usually sent to a different herbarium and can consist of more than one sheet and the same additional material (carpological, etc.) as the first specimen. If only one specimen can be made, it is called a "unicate".

Factors to consider before making/ sending duplicates

Unless a plant is suspected to be rare, very well collected in the area, or sterile, it is important to collect several duplicates (at least three if possible) for various reasons:

- The first (and most complete) duplicate must remain with the local or national herbarium of the country it was collected in

- To donate material to herbaria undertaking research on the flora in question

- To donate material to herbaria with specialists in that genus or family for accurate identification and inclusion in studies and monographs

- To ensure that the specimen is easily accessible by sending it to one or two large herbaria with good storage conditions, visiting researchers and an active digitisation programme

- To ensure the survival of the collection in case of fire, flooding, or pest damage to one or more of the other herbaria with duplicates of the collection

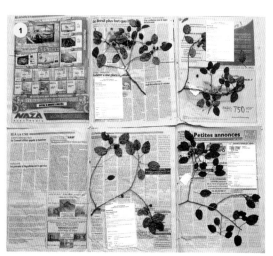

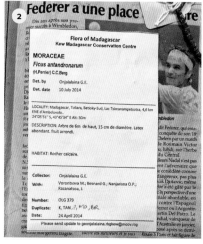

1 Several duplicates of the same specimen, with numbered tags and printed labels; ready for distribution to selected herbaria. **2** Label indicating destination of duplicates using a list of herbarium codes. **3** Duplicate with code of destination herbarium clearly noted on the newspaper.

Before collecting or distributing duplicates the following must be taken into consideration:

- Relevant agreements with institutions or laws may prohibit duplicates being sent to third parties. Researchers intending to distribute duplicates should ask for permission to do so as part of their collecting request (see: Collecting/Legislation)

- Check whether the taxon you are collecting is protected by the Convention on International Trade in Endangered Species of Wild Fauna and Flora (CITES 2021). You can do this on the Species+ (speciesplus. net website). You will need permits to import, export or re-export these taxa (living or dead and including any of its parts) if the collaborating institution in the providing country is not CITES registered

- Copies of paperwork associated with permission to visit / work in the country / area where the specimens were collected, Material Transfer Agreements and any relevant CITES permits must be sent with all duplicates

- When collecting overseas, and in the case of limited material, the first duplicate must be left in the herbarium of the country in which you are collecting. To indicate that there is only one duplicate, and to ensure that it stays in country, you should write "unicate" and the code of the national herbarium on the front of the newspaper

Selecting appropriate material
It is important that as many duplicates as possible are representative of the plant being collected. Remember that more than one sheet can be made for each duplicate. Things to look out for and

include are:

- Shade and sun/juvenile and mature leaves
- Variation in leaf/leaflet insertion, size, or shape
- Variation in leaflet number
- Variation in indumentum
- Flowering and fruiting material, immature and mature
- Both male and female flowers

In the case of potential new species, it is especially important to have good, representative material in as many duplicates as possible in case the holotype is lost and a lectotype needs to be chosen.

Labelling
Each sheet must have a tag securely attached to the collection and a collection number written clearly on the flimsy/ newspaper. If you know which herbarium it is to be sent to, write down the herbarium code, e.g., YA is the code for the National Herbarium of Cameroon (see Thiers, continuously updated).

If there is more than one sheet per duplicate, make sure that this is denoted clearly on the front of the newspaper, i.e., "YA, Sheet 1 of 2", "YA, Sheet 2 of 2", etc., so that they do not become separated and sent out as separate duplicates further on in processing. Make sure that all other material/media belonging to the same duplicate is recorded and clearly labelled with the same collection number and herbarium code.

It is important to record where duplicates are to be sent to, for each collection, and to include a list of the herbarium codes on the specimen labels before the duplicates are sent out.

COLLECTOR FEATURES
DENISE MOLMOU
Denise Molmou

Denise Molmou is a botanist at the National Herbarium of Guinea (HNG), where she organises and participates in field surveys including collecting and preserving of herbarium specimens, associated data and samples. At HNG, she also helps to identify, mount and curate herbarium specimens, and with data entry. Her interests are in documenting and conserving the native Guinean flora.

Preparing for fieldwork in Guinea
Before leaving for the field in Guinea, planning using Google Earth can save valuable time on the ground, assessing the site, the access points and where problems might occur.

You also need to present your defined objectives to the administrative authorities, for example the National Herbarium, to provide a mission order to permit travel and collection of specimens. Once obtained, you will need to present this mission order to all the local authorities and explain the context and content of your research. You need to follow the steps of the authoritative hierarchy. This can mean spending the first day going between the Regional, Prefectorial, sub-prefectorial and village councils. However, not respecting the hierarchy can cause issues further down the line.

Contacting local authorities and communities
The local authorities at the prefectural office can provide helpful advice, for example, the state of the local roads or indicate which people can facilitate your work. Members of the local communities closest to the study site who know the area well can guide you to interesting areas. They might be a hunter, a healer or someone who is used to the forest and knows the vernacular names in their local language; we try to record this where possible.

> **TIP**
> Once on site, do not forget to note all the information: the name of the site, the nearest village, the prefecture, sub-prefecture and district, and, of course, the GPS coordinates.

FURTHER READING
National Herbarium of Guinea (HNG)
www.herbierguinee.org

Ministre d'Environnement, Eaux et Forêts http://meef-guinee.org/

1 Denise Molmou at the Konkouré River, searching for Podostemaceae.
2 Recording information on local plant uses through interviews with the elders of the community in Télimélé.

SOMRAN SUDDEE
Somran Suddee

Somran Suddee is the former Head of the Taxonomy Section at the Forest Herbarium (BKF), Bangkok. He was responsible for conducting research on Plant Taxonomy and collaborating with overseas visitors on fieldwork in Thailand for the Flora of Thailand project.

Collecting in Thailand: Permits, people, and transport

It is necessary that the collector contacts the authorities to get permits before commencing collection in protected areas. For security reasons, local authority staff, e.g., the forest rangers or the army staff, must accompany the team if the expedition is in a remote area. At BKF, we normally ask the young botanists to join the trip so that they can get trained by the experienced senior botanists in the field. Four-wheel drive vehicles are always needed in areas with bad road conditions. This also needs to be arranged in advance.

Equipment

The preparation of equipment is determined by whether the expedition occurs in the wet or dry season. Generally, the equipment for collecting plant specimens in the field for BKF is similar to other institutions, e.g., field notebook, collecting pole, secateurs, GPS, binoculars, pencil, plastic bag, paper tea bag for DNA sample, silica gel, tag with collection number, newspaper, alcohol, nylon rope, etc. In the dry season, insect repellent is needed, and leech socks in the wet season.

Collecting and pressing herbarium specimens in the field

In the field, a long collecting pole is always used for tall trees or epiphytic plants on tall trees. All duplicates of a collection are tagged and put in one plastic bag. Each tag indicates the initials and surname of the first collector together with collection number and date. For a long trip, when pressing in the camp, each package of specimens is tied and placed in a bag of alcohol. All duplicates will be re-pressed and dried at the Forest Herbarium.

FURTHER READING
Forest Herbarium (BKF) https://www.dnp.go.th/botany/index_eng.html

1 Chandee Hemrat, a BKF staff technician using a long collecting pole at the Thai–Burmese border, Suan Phueng District, Ratchaburi Province. **2** Somran Suddee with BKF Herbarium staff on a collecting activity at Umphang District, Tak Province. **3** Somran Suddee, pressing specimen (with tag) in camp at Suan Phueng District, Ratchaburi Province.

ROSÂNGELA SIMÃO-BIANCHINI

Rosângela Simão-Bianchini and Fátima Otavina de Souza Buturi

Rosângela Simão-Bianchini is senior taxonomy researcher and curator at the SP Herbarium, Instituto de Pesquisas Ambientais (Institute of Environmental Research), São Paulo, Brazil. She is a specialist in the taxonomy of Convolvulaceae and Asteraceae, with decades of experience of collecting plants in Mata Atlântica and Cerrado, two very distinct biomes that characterise the vegetation of South-East Brazil.

Fieldwork in Brazil: permits, collecting and drying

In Brazil, to carry out any collection and transport of plants, you must request the appropriate permits from SISBIO and SISGEN, providing details of the collection team and the taxa to be collected in advance. Foreign researchers must register with a Brazilian team. Carrying a temporary plant press is recommended; prioritise pressing the most delicate samples. For more resistant materials, label each collection with masking tape and place them inside plastic bags for later pressing (Germán 1986; Rotta *et al.* 2008). On short expeditions, you may keep the material pressed until it reaches its destination for proper drying, but, on long expeditions, a field plant dryer is recommended. In dry and hot areas, like Caatinga and Cerrado, keep the presses under the hot sun, and they will dry naturally. In tropical forests (e.g., Amazon, Atlantic Forest), drying should be done as quickly as possible in an adequate dryer, to avoid fungal contamination.

TIP

In Cerrado, wear leggings to protect from poisonous animals, especially snakes. In Caatinga, wear a thick long-sleeved shirt; the vegetation is very high and spiny. Carry food that is easy to transport and weather-resistant, e.g., dehydrated fruits and nuts. Bring plenty of drinking water. You may freeze bottles with water and wrap them in a cloth to keep the water fresh for longer.

FURTHER READING

Instituto de Pesquisas Ambientais (SP) https://portal-sima-homologacao.azurewebsites.net/ipa/

1 Rosângela Simão-Bianchini is senior taxonomy researcher and curator at SP Herbarium (São Paulo, Brazil). **2** Making morphological observations and taking photographs in the field is important to capture key characters present in the live plants. **3** Rosângela working in the herbarium with PhD student Mayara Pastore, looking at Amazonian lianas.

THE COLLECTION OF PLANT TISSUE SAMPLES FOR DNA

Olivier Maurin, Niroshini Epitawalage, Wolf Eiserhardt, Felix Forest, Tim Fulcher, William J. Baker

Ideal plant tissues for DNA isolation can be easily collected even in the most challenging field conditions. Here, we provide summary guidance for preserving leaf or other plant parts for DNA extraction using silica gel as a desiccant (see also Maurin *et al.* 2017).

Collecting kit checklist

- Silica gel*
- Resealable bags, airtight plastic container
- Empty tea bags or coffee filters
- Jeweller's tags
- Permanent marker and pencil
- Razor blade / scalpel / surgical scissors
- Ethanol wipes
- Plant press and collecting book

Selection and preparation of fresh plant tissue

Sampling

From a single plant, harvest leaf tissue corresponding to a surface area of 5–10 cm^2. If leaves are small (e.g., ericoid leaves), sample enough leaves to reach an equivalent surface. If no leaves are available, other parts can be sampled such as leaf buds, floral parts, seeds, roots or even fresh bark.

Cleaning

Ideally, collect clean tissue, however if the tissue is dirty or shows potential contamination, clean it using an ethanol wipe or paper towel (see Fig. 1).

Preparing material

The leaf material must dry quickly to preserve the quality of the DNA. If leaves are large or very fleshy, cut them into smaller fragments to accelerate the drying process, but avoid excessive fragmentation. Bruising of plant tissue will result in enzymes being released that cause DNA degradation (see Fig. 2).

Drying out and storing the plant tissue

Method 1 (resealable bag)

One resealable bag is filled with 30–60 g of silica gel mixture*. Tissues previously prepared are placed in a tea bag or coffee filter and then transferred into the plastic bag (e.g., see Wilkie 2013). Leaves can also be directly placed into the silica gel (see Fig. 3).

Method 2 (airtight plastic container)

Tissues are placed within a tea bag or coffee filter (closed using a paper clip), which is then placed in an airtight plastic container filled with silica gel. The silica gel should at least partially cover the bag(s), which can be achieved easily by turning the box upside down (see Fig. 4).

Whichever method is used, resealable bags and airtight plastic containers must remain open for as short a time as possible so that the silica gel does not lose its desiccating properties. Silica gel indicator will change colour when fully hydrated and no longer effective.

Long-term storage

Plant material typically takes 24–48 hours to dry. It can then be transferred into individual resealable bags with a small amount of indicator silica gel, which will keep the material dry and enable monitoring of the collections (see Fig. 5).

TIPS

Sampling succulent plants
If the leaves are succulent, use a razor blade to remove epidermal slices or scoop out parenchyma tissue.

Labelling and tracking the collections
Proper labelling of the collections is critical. We recommend that each bag or filter is unambiguously labelled (with collector name and number as a minimum) and that a jeweller's tag is also included within the bag or filter. Labelling on releasable bags is done using a permanent marker when a pencil must be used on jeweller's tags.

Avoiding contamination
It is highly recommended to use clean cutting tools when sampling for plant tissue and important to clean those in between different samples.

---

*Ideal silica gel mixture: below 22A° (28–200 mesh) silica gel mixed with moisture indicating silica gel, containing a moisture indicator dye. Silica gel can be re-used, but it requires caution to avoid contamination from fragments remaining from previously collected tissue (Chase & Hills 1991).

Note the methods described here are unlikely to be sufficient for long-read sequencing of DNA or for RNA sequencing. Specialist preservation techniques are required for those purposes, which usually include freezing in liquid nitrogen or at −80°C (see Zerpa-Catanho *et al.* 2021; Yockteng *et al.* 2013).

---

1 Procedure for cleaning leaf blades if required. **2** Cut leaves in smaller fragments to accelerate the drying process. **3** Leaf fragments are transferred and mixed well with the silica gel (option 1). **4** Leaf fragments are placed within a tea bag or coffee filter, which is placed in an airtight plastic container filled with silica gel (option 2). **5** Long-term storage for tissue banking. Keep a few crystals of indicating silica in each sample to easily monitor those during storage.

THE COLLECTION OF PLANTS IN SPIRIT

Melissa Bavington

Specimens are usually preserved in spirit to enable researchers to study the three-dimensional arrangement of fleshy fruits and flowers, as there is no shrinkage (which can occur in dried material). This makes spirit material especially useful for botanical illustration.

Considerations when collecting specimens in spirit material

When collecting, the field kit should include various sizes of wide-mouthed plastic bottles, for example, Nalgene. Alternatively, to save space, resealable bags can be used, e.g., Whirl-pac or Ziploc. Double-bag or carry a plastic lunch box to ensure the safety of the specimens and to contain leaks. Using a strong local alcohol (for example, 50% ABV) is preferable to transporting chemicals. Be aware that there are specific procedures for transporting formalin and alcohol preserved collections that apply to the amount of fluid present (Huxley *et al.* 2021). It should be noted most airlines do not allow alcohol over 140 proof (70% ABV), and it is recommended that a check is also made of local rules and laws especially in countries where religious prohibitions against alcohol exist. To aid processing later, list the alcohol or chemicals used.

When preserving a specimen in alcohol rather than fixing it first, the specimen will be more brittle, and care should be taken when placing the specimen in the bottle so that it can be removed again without damage. Specimens collected domestically, for example from your Living Collections, can be placed straight into Kew Mix to be fixed (see: Herbarium Techniques/Processing spirit collections).

> **TIPS**
> - Remember to label all your collections with jeweller's tags using a spirit proof archival pen or pencil and write your collector details on both sides of the label; this way you can identify your specimen for labelling purposes without the need to open the bottle
> - Remember the specimen needs to be removed from the jar so try not to pack too tightly. The wider the bottle neck the better it is for your specimen

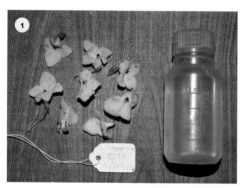

1 Example of wide-mouthed bottle used in collecting specimens for storage in alcohol. **2** Label the jeweller's tag on both sides. **3** Spirit material is especially useful for botanical artists, to observe the three-dimensions of specimens.

THE COLLECTION OF BULKY SPECIMENS

Xander van der Burgt

Plant specimens which are too bulky to press are collected and dried separately: mainly large infructescences and fruits, but also inflorescences, succulent plants, portions of stem, bark, roots, or tubers. This can also be referred to as the carpological collection.

Preparing bulky collections for preservation

When preparing bulky plant collections, a collector should ensure that those characters that can be preserved are present on the specimen. The collector and number should be attached to each part of a collection by jeweller's tags or by writing directly on the specimen. Large fruits may be collected whole to become a carpological collection but can also be sliced before drying to become an herbarium specimen. Make notes on the texture, interior tissues and colour.

Drying a bulky collection

Bulky fruit collections are not pressed during the drying process to preserve their three-dimensional structure. These collections can be placed in paper bags or mesh bags and dried in the plant dryer where warm air can freely flow around them. The drying of bulky collections can take much longer than the drying of herbarium sheets. At least every day or so, the bulky collections should be warmed up to a temperature too high for moulds and insects to develop.

Cross-referencing a bulky collection

A collector must always clearly note what they collected, both in their collection book and on their herbarium specimen labels: e.g., only carpological material, or carpological material together with an accompanying herbarium sheet. This will help the curation process where bulky collections such as fruits often become separated from the accompanying herbarium sheets.

FURTHER READING
Letouzey (1986).

1 Herbarium specimens and bulky collections are dried with an electric fan heater. **2** Herbarium specimens and bulky collections are dried on a portable dryer using several camping gas burners. **3** Large fruits can also be sliced and then dried.

THE COLLECTION OF ETHNOBOTANICAL OBJECTS
Mark Nesbitt and Ben Hill

Human interactions with plants are widely recognised as central to botanical research into plant-based livelihoods, conservation and biocultural significance. Alongside field collection of uses data through notes and images, collection of plant and fungal raw materials and products creates a valuable resource for research and public engagement (Salick *et al.* 2014).

Definition and purpose
Ethnobotanical specimens are very different to herbarium specimens, coming in all shapes and sizes, and with a focus on attributes that reflect use rather than taxonomic identification. A random collection of plant-based objects will not be of research value. These principles will lead to a useful resource:

- Collecting the sequence of raw material, part-finished object and tools, and completed object is most useful in recording and transmitting how plants are being used
- As far as possible, both botanical and vernacular names should be recorded for all components of ethnobotanical specimens
- Objects chosen should represent a coherent story, for example, of a traditional or new plant industry, sustainability, or Fairtrade

Ethical and legal aspects
These will depend on the collecting context. If this is a supermarket at home, or the purchase of a basket at a market overseas, these are unlikely to involve formalities. Any work that involves detailed research into traditional knowledge (TK) raises ethical and legal matters of consent, fair reward for participants, and compliance with local regulations and with relevant international treaties (see: Legislation). Particular care is required when medicinal plants are the main topic of research owing to widespread concern over unethical bioprospecting. Good practice in benefit-sharing includes deposition of sets of ethnobotanical specimens in museums in the study region and co-authorship of publications by local experts.

Collecting methods
Ethnobotanical specimens can be collected anywhere that plants are being used, whether a shop, market, factory, or household. Objects that are no longer in use can be made to special order if the expert knowledge and materials are still locally accessible, or are sometimes available secondhand. However, more ambitious studies that link ethnobotanical specimens back to the harvested plant are of most benefit and enable the collection of herbarium voucher specimens to confirm plant identity. Most formats of specimen can be collected – for example, basketry, textiles, paper, fibres and dyes, foods, poisons, crude drugs, and other items used in daily life – so long as they are stable in dry or liquid form. Collecting numbers, notes and field photographs should be implemented as with herbarium specimens (Bye 1984; Alexiades 1996; Martin 2004).

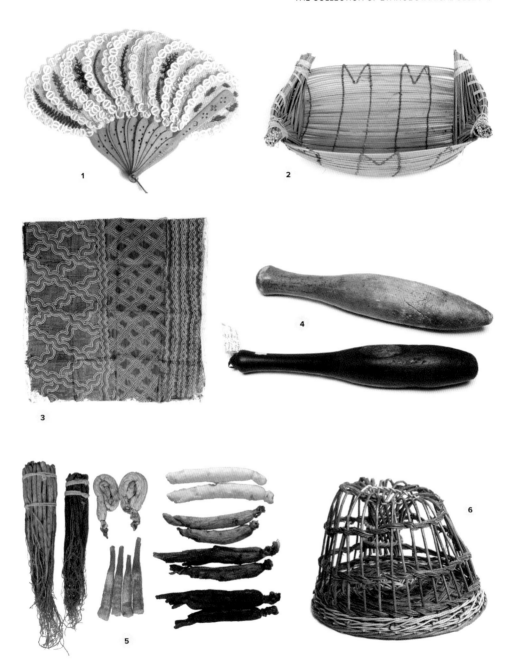

Recent acquisitions to Kew's Economic Botany Collection.

1 Lacebark fan (*Lagetta lagetto*) from Jamaica, c. 1880. Gift in 2019 from Marion Maule.
2 Fish-trapping basket, Madagascar, midribs of *Raphia* leaves, collected by Lauren Gardiner, 2016.
3 Resist dyeing in progress, using wheat paste, The Gambia, collected by Jenny Balfour-Paul, 1997.
4 Indigo beaters, Senegal, collected by Jenny Balfour-Paul, 1997.
5 Roots of ginseng (*Panax ginseng*), China, collected by Christine Leon and Lin Yu-Lin, early 21st century.
6 Lobster pot, made in Great Yarmouth from willow (*Salix* sp.), collected by Maurice Bichard, 1991.

THE COLLECTION OF SEEDS

Daniel Cahen

Seed banks are a cost-effective tool for the long-term *ex situ* conservation of plants, although how seeds are collected and processed affects their longevity. Unless the seeds are identified, they have little scientific value. Voucher herbarium specimens are therefore of critical importance because they are used to identify seed collections (ENSCONET 2009).

Seed collecting

Seeds are best collected when mature, from plants in the process of natural dispersal. It is recommended to collect at least 10,000 seeds from at least 50 plants sampled randomly and evenly across a population. No more than 20% of the mature seeds available that day should be collected in order not to threaten the population. If the proportion of empty, infested, immature, or aborted seeds is greater than 30%, collectors should seek another population or increase the number of collected seeds (Way & Gold 2014).

Seeds need to be dried carefully to increase their longevity. Spreading them thinly in a shaded, well-ventilated space or using a desiccant may be necessary. The flesh of over-ripe fruits can be removed using a sieve and water (Gold 2014).

Herbarium voucher specimens for seed collections

Seedbank herbarium vouchers should represent the plant population the seeds were collected in (Jennings *et al.* 2018). It will usually be most useful to collect specimens that reflect the average morphology of the plants while writing the full range of variability.

Flowering specimens are often necessary to identify vouchers, and flowering parts should be collected when possible. Collecting sterile specimens should be avoided because these often cannot be confidently identified.

The same collection number should be used for both the seed collection and the associated herbarium vouchers to easily cross-reference them. It is recommended to collect herbarium vouchers for every institution that will receive seeds but also duplicates to send to other institutions because they are valuable additions to herbarium collections in their own right.

If no voucher can be collected, it might be possible to regrow a cultivated voucher from the seeds, especially for herbaceous plants with a short life-cycle. When a specimen cannot be collected and a cultivated voucher cannot be regrown, it is best to have the plant identified in the field by an acknowledged expert whose name is recorded and to take photographs that can be linked to the seed collection.

1 Unripe *Iris* capsules; too early for a seed collection. **2** *Iris* seeds in the process of natural dispersal; optimal time for a collection. **3** Cut test to check whether *Iris* seeds are empty, infested, immature or aborted. **4** Post-harvest handling of seeds.

PHOTOGRAPHY

Daniel Cahen

Photos complement herbarium specimens by showing features that are lost on dried specimens, such as a plant's habit and flower colours. They can also be used to document plant occurrences when collecting a physical specimen is not possible (Gómez-Bellver *et al.* 2019). Scientists benefit from images that are clear and informative and that are made available online on public repositories (Struwe & Nitzsche 2020).

1 *Euphorbia characias* cyathia close-up (clear image). **2** *Euphorbia characias* whole plant (clear image). **3** Compositae cypsela and pappus details can be diagnostic for identification (informative image). **4** Brome spikelet on a ruler (informative image).

Photography technique for biological records

Images documenting a plant occurrence generally must meet the following criteria:

- *Composition:* what the subject is should be clear (whole plant, flower, etc.) and kept in focus as much as possible. It is usually in the centre of the image against a contrasting background
- *Exposure:* informative characters can be lost if the subject is not properly exposed. It is helpful to use a flash diffuser for photos taken very close

Close-ups are often essential for identification. Cameras with a macro lens give the best results, although phone cameras can take increasingly clear images.

Taking informative pictures

Knowledge of useful identification characters helps one take more informative photos, such as including images of phyllaries and the fruit pappus in Compositae. As a rule of thumb, it is good to take multiple photos, including:

- The whole plant in its habitat; it may be necessary to manually focus on the plant
- Leaves and stipules when they are present
- Flower close-up, front and side view
- Fruit

When attempting an identification in the field, it is useful to take photos while progressing through the identification key and to include a ruler in photos where measurements are needed. Visible diagnostic characters make it easier for others to identify observations.

Making pictures available to scientists

Making images available online, together with their date and geographic coordinates, not only helps in understanding the distribution of species but also gives phenological and morphological information. These photos can be reused by others in a variety of publications if accompanied by a non-restrictive copyright licence, such as CC-BY or CC-BY-NC.

The largest global source of biological occurrence data is the Global Biodiversity Information Facility (GBIF), which stores biodiversity data from hundreds of organisations. Data exports from GBIF are widely used in scientific research, and making photos available on the platform is an effective way for them to be accessible to scientists.

iNaturalist, a citizen science social network for naturalists, is an excellent resource for sharing images of plant occurrences. A computer vision model can help identify observations, which can then be exported or filtered by taxonomy, location, or date. An active user community aids in the identification of observations, and these are made available on GBIF when more than two-thirds of users agree on the identification (Boone & Basille 2019).

HERBARIUM HIGHLIGHT:
EAST AFRICA HERBARIUM (EA)

Paul Musili and Kennedy Matheka

The East Africa Herbarium (EA) is one of the sections in the Botany Department of the National Museums of Kenya (NMK) in Nairobi. EA is a national and regional repository and reference centre for plants and fungi, with the collections acquired through expeditions and collaborative programmes over the last 120 years.

The East Africa Herbarium. **1** Botany building main entrance. **2** Botany building rear elevation. **3** Internal view of the Herbarium showing the cupboards and digitisation and imaging set up. **4** The Botanical Library at EA. **5** Staff and students undertaking a curation and training session inside EA.

History and aims

EA was established as the Amani Institute Herbarium in Tanzania, in 1902 and later moved to Nairobi in 1950; it was incorporated as herbarium in the Coryndon Memorial Museum (later the NMK) in 1958 (Newton 2004).

With about 60 current staff, it undertakes taxonomic research, field surveys, specimen identification and name verification (e.g., including for the Kenya Wildlife Service, Kenya Forest Service and local Universities), curates a comprehensive databank, and has an active training programme.

Coverage and scope

Size and scope

The EA Herbarium has approximately 1.2 million specimens, comprising mostly vascular plants with a few bryological (c. 1500) and mycological collections (c. 3000 Ascomycetes and c. 2000 Basidiomycetes); the Nairobi Botanic Garden is a complementary part of the Herbarium. EA's mission is: '*To collect, preserve research and disseminate botanical information on Kenya and the East African region for conservation and utilization*'.

Arrangement

Families are stored systematically with dicots following the Hutchinson classification, while monocots follow Dahlgren's system. The species within families are arranged systematically according to the Flora of Tropical East Africa (FTEA). However, species name changes are updated following the APG system.

Research and collections management

EA staff undertake botanical surveys throughout the country culminating in approximately 2,000 specimens per year; with several new species to science being described.

EA uses BRAHMS software for their database and more than 300,000 specimens have been databased (Pearce *et al*. 1996). Additionally, specimen digitisation is ongoing; and nearly 4,000 specimen records, primarily types, are available on JSTOR Global Plants. Recently, EA has digitised all the botanical illustrations, photographs and slides (c. 8000 items) and are currently digitising the 150k species data cards.

The collections are monitored for pests using light traps, and any infestation is controlled by deep-freezing and annual fumigation.

Conservation programmes

EA staff also contribute living collections of rare or endangered species to the Nairobi Botanic Garden for *ex situ* conservation. In addition, EA staff participate in collecting and identifying the many orchid species that are conserved within NMK as living specimens.

EA staff work closely with NMK's Coastal Forests Conservation Unit (CFCU) on the Kenyan coastal flora, including over 50 sacred forests, 'Kayas', that have been gazetted as national monuments.

FURTHER READING

NMK Botany Department http://museums.or.ke/botany-department/

EA specimens on JSTOR https://plants.jstor.org/partner/EA

Nairobi Botanic Garden https://tools.bgci.org/garden.php?id=1375

Flora of Cameroon

Thymelaeaceae.
Dicranolepis sp.
Det. D. Thomas October 1993

Cameroon. East Province. Boumba-Ngoko Departm
Moloundou Arrondissement; Lopondji village. Und
forest. Shrub; large white showy flowers on stem; 1
style; 5 sepals; 10 petals; calyx green underneath.

Vernacular Name: Ngbi (Baka).

Jefferson Hall 016/92   29 Nov

HERBARIUM TECHNIQUES

ACQUISITIONS AND DISPOSAL
INTRODUCTION
Lauren Phelan and Alan Paton

This section outlines the factors that impact the decision to acquire material including institutional policy, agreements, and relevant legislation.

Policy

Herbaria need to establish policies around what material should be acquired, appraised, deaccessioned or disposed of. These policies will be influenced by the strategic aim of the institution, for example the current priorities for collection and the long-term goals of the institution. Such policies are often called Collection Development Policies or Plans (Huxley *et al.* 2021). An example template for a Collections Development Policy is provided by the Arts Council England (2014), which highlights the areas that should be considered in creating such a policy. Standardising policies and processes across the collection helps embed best practice and assists compliance with relevant regulations and laws.

> **TIP**
>
> An herbarium can become registered with CITES. Being a CITES-registered institution helps facilitate movement of specimens for scientific study between registered institutions. Registration requires discussion with the relevant national CITES Management Authority. If sending CITES-listed specimens from an unregistered institution, there must be cooperation with the recipient prior to sending a shipment. This allows time to apply for the export and import permits to ensure specimens are received legally.

Legislation

Collections must ensure that acquired material is collected in accordance with relevant national law. Laws on access to genetic resources and benefit-sharing may impose conditions on material such as restrictions on supply to third parties or sharing data on localities or traditional knowledge captured with the specimen. Procedures should be in place to ensure collections are treated in accordance with the terms and conditions under which they were supplied. Potential acquisitions may also be covered by legislation implementing the Convention on International Trade in Endangered Species (CITES), so material being sent from other countries needs to have the appropriate CITES permits. Material may also be subject to national laws and processes on plant health or regulations governing collection from protected areas.

I hereby donate the material and associated data to Kew to use for the stated purposes and declare, that to the best of my knowledge and belief, the information that I have given in this document is accurate.

Signed:

Name:

Position/institution/full address:

Date:

Scientific name(s) or description of material if greater than 20 specimens	Wild collected/ natural source? (Y/N)	Material type - herbarium sheet, carpological, DNA, plant, cuttings, seeds, in vitro culture, etc.

Additional terms and conditions of use of material (from original permits or access agreements)	
Field or database records attached (yes/no?)	
Permits/ documentation attached, (please explain where not). e.g. collecting permit, export permit, CITES label etc	

FOR KEW INTERNAL USE ONLY

Material received at Kew/ Wakehurst by:	Staff name: Department: Email/telephone:

Part of the Kew standard donation letter detailing the information needed to be completed by the donor.

Donation

The donor of material may also wish to impose restrictions on use of material. Material should only be accepted if any terms and conditions can be met and if the material can be used to fulfil the purpose of the collection. It is useful to have a donation letter in which the donor provides evidence of legal collection and agrees to a set of standard uses which are permitted for the material being donated.

SPECIMEN ENTRY, ACQUISITION, AND ACCESSION
Lauren Phelan and Alan Paton

This section describes the processes around receiving specimens and accepting them into the permanent collection.

Receiving incoming material

All incoming consignments of specimens should arrive at a single point of entry to the collection and be taken to a secure area isolated from the rest of the herbarium. This area should be considered 'dirty' or 'quarantined' due to the risk of pest infestation. Here, the contents of parcels can be checked, inspected for any signs of damage during transit, and the presence or absence of paperwork or required permits can be noted. The consignments then need to be treated for any potential pest infestations by either deep freezing or warm air processes (see: Pest management; Biosecurity).

Consignments may be contaminated with chemical pesticides such as naphthalene or camphor. The use of these should be discouraged as they may be toxic. The pesticides need to be removed in a fume cupboard, while wearing appropriate personal protective equipment, and any packaging disposed of as hazardous waste (Makos *et al.* 2019). After the potential pest treatments are complete, the parcels should be moved to a 'clean' area in the main herbarium building where they can be processed, and acknowledgement of receipt can be made to the sender.

1 Walk-in freezers used to deep-freeze specimens with potential pest infestations. **2** Loading the chest freezer. **3** Consignment label, showing summary details. **4** A consignment of specimens received damaged in transit.

Recording details

Consignments of specimens may be received from staff on fieldwork, other institutions or private individuals. Incoming material should be recorded upon entry so the specimens can be accounted for at all times while in your care. The information that should be recorded includes:

- Details of the sender/collector
- Date of arrival
- Status, e.g., donation, exchange, purchase, field collections from staff, visitor's own material temporarily in the herbarium (see: Herbarium techniques/Loans)
- Any timelines associated with the incoming material (e.g., identifications needed by a certain time)
- A list and number of specimens, named to family/genus if possible, or a list of collection, numbers, as a minimum
- Country of origin of all specimens
- Details of the evidence for legal collection, and any terms, conditions or restrictions regarding use of the material such as on permits or agreements

The material should be clearly labelled with a summary description, action to be taken and provided with an identifier cross-referencing the material to the information received from the provider.

Evaluation

Incoming material needs to be evaluated to enable a decision on whether to keep the material or dispose of it; staff members responsible for overseeing the process should be identified. The following factors should be considered when making the decision:

- Does the material fit the strategic aims and focus of the herbarium collection?
- Does the material add value to the collection, e.g., is the material filling gaps in the current collection or duplicating already well-represented areas or species?
- Are the terms and conditions attached to the material clear; can any restrictions placed on the material by the provider be implemented effectively and efficiently over time? If legal status is unclear contact the donor for clarification and consider returning the specimens) (see: Record Keeping)

If the specimens are accepted and accessioned into the permanent collection then each specimen should be recorded, preferably by marking with a unique identifier such as a barcode or accession number; or by some other method of cross referencing the specimens back to the original details of their provision.

> **TIP**
> In traditional museum collections the term 'acquisition' means that ownership has been transferred to the collection and 'accessioning' means it has been formally incorporated into the permanent collection (Collections Trust 2022). However, specimens in natural history collections may also fall under the scope of the access and benefit-sharing regime of the Convention on Biological Diversity, and the provider country may assert their right through material transfer agreements to permanently own collections provided from that country. An herbarium need to ensure that it records whether it owns material incorporated into the collection, or ownership remains with the provider.

FURTHER READING
Collections Trust (2022; Spectrum 5.1: Acquisition and accessioning).

DISPOSAL AND DEACCESSIONING
Lauren Phelan and Alan Paton

Processes related to the disposal of collections and deaccessioning of collections previously accepted into the permanent collection.

Plant waste bin for biosecure disposal. Material should be incinerated and not disposed of in general waste.

Introduction

An herbarium should develop a clear policy and processes for disposing of specimens and the staff members with authority to make disposal decisions should be identified. The decision to dispose of a specimen will be guided by several principles:

- The strategic goals of the institution
- The fit to the scope of the collection and its priorities
- Any constraints imposed on the herbarium by its governing principles
- Any agreed conditions surrounding particular collections set at the time of acceptance

The disposal of specimens enables better care for the collection as a whole and should not be for financial gain.

Methods of disposal

Specimens may be disposed of before formal accessioning into the collection or afterwards. In either case a record of the disposal should be made. Incoming specimens that do not fit the remit of the herbarium may be returned to the donor or passed to another collection if the terms and conditions of acquisition allow such transfer. Herbaria have a long history of exchanging duplicates to improve their collections, facilitating expert identification and providing a mitigation against loss due to disaster. Incoming specimens of poor quality, for example without provenance information, can be destroyed if their conditions of supply allow. If such specimens originated from other countries they should be incinerated, rather than being put in the general waste to reduce the risk of introducing invasive species, pests and disease.

Deaccession

If a specimen has been accessioned, the herbarium has made a long-term commitment to care for it. The need for deaccession may arise for strategic reasons such as a change of research focus; legal reasons could include the discovery that the herbarium does not have legal title to the specimen, and it really belongs to another herbarium. The reason for deaccession, and the evidence that the herbarium is able to dispose of the material under the terms and conditions of its initial acceptance, should be documented and approved by the herbarium governance. Decisions around deaccessioning should not be taken by an individual. Ideally, deaccessioned specimens should be offered to another herbarium. These may have been previously cited by researchers as being found in the herbarium, so it is important to record the deaccessioning within the herbarium catalogue with details of where the specimen is now housed.

FURTHER READING
Museum Association (2014); Collections Trust (2022; Spectrum 5.1: Deaccessioning and disposal).

SPECIMEN EXIT AND SHIPPING

Lauren Phelan and Alan Paton

The processes around specimens being sent from herbaria and their shipping.

Specimen exit

Specimens or other material such as samples may leave the herbarium through sending and returning loan transactions, gift or exchanges to other institutions or disposals such as returning specimens to their donors if not accepted into the herbarium collection. Specimens leaving the herbarium need to be recorded and staff who can authorise the specimens' exit identified. In all cases the material should be sent out with documentation describing the material and explaining why the material has been sent. Always provide material transfer or supply agreements setting out how the material can be used by the recipient when sending accessioned plant material from the permanent collection.

Shipping

Specimens should be securely packed so they do not get damaged in transit. Boxes should be wrapped with moisture- and water-resistant paper, such as medium-heavy duty Kraft Union, which will help protect against water damage during transit. Parcels sent by sea may take many months to arrive and extra protection during transit is necessary.

Shipping forms, specimen lists and tracking numbers may now be sent and received electronically, but specimen parcels must always have a physical copy enclosed with them. Shipping forms should be duplicated as necessary, with one copy being retained by the dispatching institution and another enclosed with the parcel ready for the recipient to sign and return as confirmation of safe receipt. In addition to this another copy should be sent separately and in advance of the shipment to alert the institution to expect an arrival of specimens, and in the event of non-delivery the recipient will be able to inform the sending institute that it has not arrived. The shipping form should include the status of the specimens in the consignment (e.g., loan, exchange, gift, return of loan) as well as a description of the specimens and total, or a complete list of specimens.

Shipping regulations

Special attention needs to be paid when sending specimens to countries with strict plant health or biosecurity procedures (such as Australia and New Zealand) and advice must be sought from the receiving institute prior to sending any shipments. The use of phytosanitary certificates is becoming more common, and the movement of material covered under plant health legislation may require specific packing methods and an accompanying Letter of Authority to move or import material.

Legislation and shipping requirements will change over time, vary between shipping agents, and by country to country — it is always advisable to contact the receiving institution before sending a shipment, thus avoiding the unnecessary delay or return of a consignment.

Once a consignment has been dispatched the location record of the specimens should be updated.

1 & 2 Before and after packing a consignment of specimens.

FURTHER READING
Collections Trust (2022; Spectrum 5.1: Object exit).

RECORD KEEPING
Lauren Phelan and Alan Paton

This chapter outlines the importance of record keeping in different aspects of herbarium management.

Introduction

Herbaria should record the movement of specimens in and out of their collection. Details of incoming specimens should be recorded with a description of what has arrived, the reason for it being sent, such as borrowed material, returning loans, enquiries, or acquisitions; and the terms and conditions attached to incoming material which govern the subsequent use of the collections (CETAF 2019). Within the collection, records may need to be made of where specimens or consignments are kept during their journey from arrival to eventual incorporation into the collection or if they are temporarily moved, for example to be digitised, so that the specimen or consignment of specimens can be found easily. Specimens leaving the collection for loan, sampling, disposal or deaccession should also be recorded, along with the terms and conditions of supply to receiving institutions. Records should be kept and filed indefinitely, as they may need to be referred to in the future and could be of historical significance. This chapter considers the information that should be captured when specimens arrive at, or leave, the herbarium. Computerised collection management systems are briefly introduced and some useful resources provided.

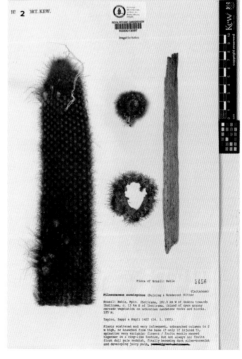

1 Recording consignment details on a collections management system. **2** A unique identifier (barcode) on an herbarium specimen sheet. **3** Paper consignment records.

Records of incoming Specimens

Incoming material should be recorded as soon as possible after its arrival. A record for each consignment of incoming specimens should be created with a unique identifier, or in the case of returning loans a link made with the outgoing loan record. Ideally records should be made in a computer system, but could be made in a ledger or paper files. The unique identifier for the consignment allows information to be retrieved easily and helps to avoid specimens from different consignments being mixed up. It also provides a reference for linking individual specimens to the detailed information about the consignment. The unique consignment identifier should be issued in chronological order of entry to facilitate retrieval, particularly in non-digital recording systems.

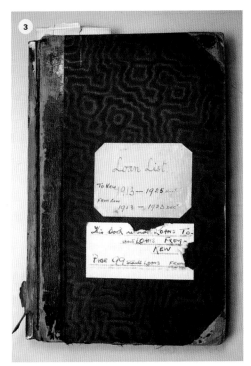

The following information about the consignment should be recorded:

- Type of consignment e.g., loan, acquisition, enquiry
- Brief description and number of specimens; databased records of individual specimens if available
- Sender's reference number (if applicable)
- Sender's name and address
- Date of arrival
- Due date (if a loan)
- Country of origin of specimens
- Proof of legal collection/donation or supply, e.g., donation letter or material transfer agreement
- Any restrictions on the use of specimens, which must be linked to specimens and filed for future reference
- CITES label/permit numbers if appropriate as these may need to be reported to the relevant CITES authority
- Condition of specimens upon arrival, noting any damage, pests or risks, e.g., packaged with naphthalene

Documents such as permits, material transfer agreements, and the sending institution's letters or instructions related to the consignment should be filed and linked to the consignment record containing the information above. The receipt of the consignment should be acknowledged so that the sender knows it has arrived safely.

Incoming specimens, whether kept in bundles or as individual specimens, need to be marked with the consignment identifier so that the specimens themselves can be linked to the relevant consignment details (see Fig. 3 page 76). Digitising individual specimens and attaching a unique specimen identifier such as a barcode or accession number allows easier tracking and linking of the individual specimen record to the consignment record. Unique specimen identifiers also allow linkage of specimen data to other information about the specimen and outputs arising from the use of the specimen (Lendemer *et al.* 2020). Specimens can be digitised on arrival or after study and decision to keep the material, depending on the nature of the transaction. If no computerised system is in place, the consignment identifier can be marked on each sheet, or the specimen given a label which summarises consignment information most relevant to the use of the specimen.

It may be necessary to contact the provider of the specimens if the consignment lacks the appropriate documentation such as a donation letter, permit or material transfer agreement which outlines how the material may be used.

> **TIP**
> Spectrum is a collections management standard developed by the Collections Trust (2022; Spectrum 5.1: Introduction to Spectrum), and provides a useful overview of the general collection management processes describing the important information which should be recorded.

For collections falling in scope of national legislation to implement the Nagoya Protocol, research resulting in new insights into characteristics of the genetic resource which is of (potential) benefit to product development may need to be reported through the relevant national competent authority. Details can be found from the ABS Clearing House (see Glossary).

National laws safeguarding personal data may apply to institutions holding collections. Generally personal information needs consent to be used and herbaria should inform people whose data they collect, and that these data may be used for relevant collection management purposes. Herbaria should not use such data for other purposes.

Records of outgoing specimens

Herbaria should record the movement of consignments of specimens leaving the herbarium for whatever reason, for example, loan, returning borrowed specimens, gift, exchange, disposal or deaccession. These should be recorded and retrievable for future reference or if any problems arise with their shipping. As with incoming specimens the outgoing consignment should be given a unique consignment identifier either on a computer system, paper file or ledger. A material supply or transfer agreement which explains the sending herbarium's conditions of use should be prepared and sent with the consignment. The agreement should also note any restrictions on the use of specimens resulting from the terms and conditions under which they were received, which need to be followed. The following information should be recorded for outgoing consignments;

- Type of consignment, e.g., loan, gift, disposal etc.
- A copy of the material supply or transfer agreement
- Brief description and number of specimens: databased records of individual specimens if available
- Your reference number
- Recipient's name and address
- Date of sending
- Due date (if a loan)

- Country of origin of specimens
- CITES label/permit numbers if appropriate as these may need to be reported to the relevant CITES authority

A copy of the consignment record should be sent to the recipient to sign and return as acknowledgement of receipt. It is helpful to send a copy under separate cover (via email or post), alerting the recipient that the shipment is coming and of any tracking numbers in case any parcels go missing or there are other shipment problems. The receipt from the recipient should be filed, and the consignment record updated, noting any discrepancies or comments from recipient.

Some countries require imported specimens to be accompanied by phytosanitary certificates or other biosecurity permits (for example, Australia: BICON – Import Conditions – Conditions (agriculture.gov.au)). Recipient collections should be contacted in advance to clarify any necessary biosecurity procedures.

Methods of recording

There are many different collections management systems ranging from open-source to proprietary software. Typically, such systems allow tracking of the movement of specimens, their terms and conditions of use, and can link the specimen data to images and relevant documentation related to specimen transactions such as loans. Managers wanting to move to or update a computerised system should carefully consider their current and future requirements and select a system which matches their requirements and available budget. They also need to take into account the level of technical skills necessary to run the system and what staff are available to administer

and be responsible for the system. The options available will evolve over time and herbaria should consider a wide range of options when making a choice. Some useful resources are listed below, but the market changes over time and the resources may not be updated. For example, at the time of writing, the Royal Botanic Gardens, Kew has chosen Earthcape (https://earthcape.com/) for their science integrated collections management system. However, this system has only recently been added to some of these resources mentioned in Further Reading below.

If a computerised system is unavailable, documentation should be kept as a physical archive and be retrievable for future reference. Where possible, digital scans of correspondence and accompanying permits and paperwork should be made to allow quick retrieval.

FURTHER READING

iDigBio: Biological Collections Databases, Tools, and Data Publication Portals. https://www.idigbio.org/wiki/index.php/Collections_Management_Systems

Collections Trust. https://collectionstrust.org.uk/software/

Software for Biological Collection Management – TDWG Subgroup on Accession Data. https://www.bgbm.org/TDWG/acc/Software.htm

Museums & Galleries of New South Wales (MGNSW), Collection Management Systems. https://mgnsw.org.au/sector/resources/online-resources/collection-management/collection-management-systems/

Best Collections Management Software in 2022: Compare 60+ | G2. https://www.g2.com/categories/collections-management

THE EXTENDED SPECIMEN CONCEPT

J. Mason Heberling and Bonnie L. Isaac

The extended specimen concept considers herbaria not as isolated static collections of individual specimens in the traditional sense, but rather as diverse streams of globally interconnected digital and physical sources of information. Herbarium use is always evolving and so must our approaches for collecting, curating and engaging with specimens.

Extending the relevance of herbaria

Specimens are more than meet the eye. Bolstered by digitisation, technology, and new perspectives, herbaria are receiving unprecedented attention across the sciences and humanities. Specimen use has diversified from a few topics to the many today, ushering herbaria into a new era of scientific, educational, and societal relevance (Heberling *et al*. 2019). Specimens are increasingly used in ways unanticipated by botanists a century ago. The extended specimen movement seeks to integrate otherwise disconnected, multi-faceted data to facilitate this new wave of specimen use (Webster 2017).

Maximising the potential of collections

Information in the world's herbaria is vast, with much to discover. Each specimen's digital record is continually enriched with taxonomic, genotypic, phenotypic, chemical, and environmental, and historical data. The digitised specimen record serves as the primary extension. Secondary extensions include data derived from the specimen and ancillary material from the collection event (e.g., field images). Tertiary extensions include coarser-scale information captured by multiple specimens and sources, such as monographs or trait databases (Lendemer *et al*. 2020).

Curating new data alongside physical specimens poses a massive challenge for herbaria, e.g., to keep up to date with new technologies and formats, as well as the time required to manage their large size (often continually growing). Workflows enable efficient, large-scale data extraction from specimens, including artificial intelligence-based image analysis and crowd-sourced digitisation. New data should link collection databases to online repositories of genetic, phenotypic, phenological, taxonomic, and ecological data. Researchers should cite digital object identifiers (DOI), such as those provided by the Global Biodiversity Information Facility (GBIF), to link data and publications back to specimens (Heberling *et al*. 2021).

Rethinking the specimen

Nature is more than fits on an herbarium sheet. Collection practices must adapt to enable the next generation of use, including re-evaluations of how and what we collect. New collections can be "born extended" with broad uses in mind, such as downstream digital analyses and demand for destructive sampling. Community science platforms, such as iNaturalist, can be leveraged to facilitate data collection in the field and collect observational data to complement physical specimens (Heberling & Isaac 2018).

CARNEGIE MUSEUM
HERBARIUM (CM)
PITTSBURGH, PA.

N° 535212

PLANTS OF PENNSYLVANIA

MELANTHIACEAE

Trillium erectum

Identified by: B. L. Isaac 2018

United States: Pennsylvania: Lawrence County: 7.3 mi
WNW of Prospect, along Slippery Rock Creek, McConnells
Mill State Park, wooded slope along floodplain. Flowers
purple brown, population has yellow flowered individuals
and some with deflexed peduncles. Lat/Long: 40.9389, -
80.17785 WGS84

Bonnie L. Isaac 25050   5 May 2018
    with Joseph A. Isaac

HERBARIUM, CARNEGIE MUSEUM (CM), PITTSBURGH, PA

A single herbarium sheet is the core of the 'extended specimen concept' and is linked to limitless databases.

1 Physical herbarium sheet (Isaac 25050; CM535212).
2 Online digital record (midatlanticherbaria.org).
3 iNaturalist record, linked to QR code on label.
4 Trait data archived in trait databases (e.g., try-db.org).
5 Environmental data at time of collection.
6 Genetics data archived in databases (e.g., Genbank/NCBI).
7 Archival and library materials digitised online.
8 Species associations (e.g., dispersers, root fungi).
9 Related collections (e.g., spirit, pollinators).
10 Biogeographic data aggregated from data networks (e.g., gbif.org). Adapted from Webster (2017).

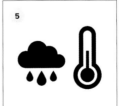

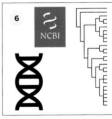

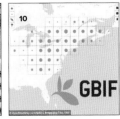

BOTANICAL ILLUSTRATION

Christabel King

Botanical illustrations are used in publications, journals and Floras or in any situation where a visual image can clarify and augment a written description. This section introduces the reader to botanical illustration and gives some guidelines to making illustrations; detailed advice can be found in King (2022).

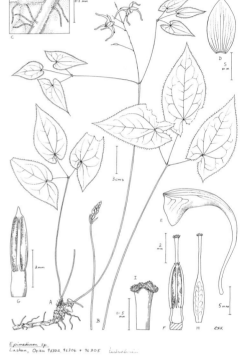

Epimedium lischichenii, specimen and illustration.

Introduction

A line drawing illustration is often preferable to a photograph, which can show too much detail irrelevant to taxonomic diagnosis. In a hand-drawn illustration, however, objects can be shown together at different magnifications and there is no distracting background. A simple outline drawing may be all that is necessary to show the important characters, but both photographs and line drawings can be included to complement each other (e.g., Flora of Thailand and Flora of Peninsular Malaysia).

In practice, botanical illustrations range from simple outline drawings to highly detailed naturalistic images, and they may consist of many separate details drawn at different magnifications. Whatever the style of drawing they need to be clear and accurate and show the diagnostic features, including those important for identification.

Stages in making a botanical illustration

- Select suitable specimen(s) for illustration

- Find the required dimensions for the illustration according to where it will be published and what reduction will be used in printing

- Decide what the drawing should include. The commissioning botanist should provide a list of features to be illustrated

- Design the illustration in pencil and show it to the commissioning botanist for approval

- Make the illustration

Specimen and feature selection:
Typically, a line drawing will include a portion of the plant at the flowering or fruiting stage to show the habit. This will be a carefully chosen example specimen from which can be seen the main features of phyllotaxis, leaf shape and type of inflorescence. Details of flowers, fruit and any other important features can be arranged around the habit. The commissioning botanist chooses which specimen to illustrate and also provides a list of the parts to be drawn and which need magnified details.

To get an idea of the sort of illustration needed it is worth looking in peer-reviewed international journals, for example *Kew Bulletin*, *Edinburgh Journal of Botany and Phytotaxa*, for recently published illustrations of related taxa. Features important to illustrate vary widely among different taxonomic groups.

Dimensions and reductions:
A reduction of x ⅔ in printing is widely used in journals but sometimes a half (x ½) is used.

Designing the illustration

A sketchbook with smooth surface cartridge paper is useful for botanical and microscope drawings. A spiral-bound A4 size is convenient as the cover can be turned right back so that it is not too big when open on the desk. Begin the drawing with an HB pencil and then increase the precision with a sharp 2H pencil for accurate detail. Keep the point sharp by rubbing on emery paper. Good-quality cartridge paper will withstand a certain amount of erasing and re-drawing.

Remember to record the name of the plant drawn, the date the drawing was made and any specimen data such as collector's name and number, barcode or herbarium accession number (especially if duplicates are available) and when and where it was collected. Even if it is not possible to make a finished illustration immediately, a careful drawing in a sketchbook with information about when and where it was collected is a useful record of a plant and may provide the basis for a finished illustration at a later date.

> **TIPS**
>
> **Drawing dissected plant material:**
> Information about making floral dissections is to be found in a separate section. Drawings can be made in a sketchbook as above. Please see the paragraph on Magnification below for advice on choosing a suitable size for an enlarged detail.
>
> **Camera lucida:** This is an extension to a microscope which superimposes an image seen through the objective on a drawing. The eye can then follow the point of a pencil as it makes a drawing of the image on the page. The advantage of this is that an accurate drawing can be made without artistic ability. Some neatening of the lines may be necessary to make it a sufficient basis for an illustration.

Magnification

The purpose of a magnified detail drawing is to make visible a feature that cannot be clearly seen at natural size. The size at which it is drawn should take into account any reduction at the printing stage. A reduction of ⅔ the size of the original is often used but half size is sometimes used. Details should be drawn large enough to still be clear when reduced. If a detail is drawn too small, fine lines may coalesce and shading become patchy when printed. Make sure the size will fit into the layout of the whole illustration and note immediately at what scale it

has been drawn because it is easy to forget. To inform the viewer how big something is when printed, a scale bar of a stated size can be drawn beside each object on the artwork. A caption will give the name of the part drawn and the size of the scale bar or just the magnification, for example, 'A, Petal, scale bar 10 mm' or just 'A, Petal, x 2'. In the second case one must be careful that the printing reduction has been taken into account, so if, for example, the drawing will be printed ⅔ of the original size and the caption reads 'Petal, x 2', the petal must be drawn x 3 on the original artwork.

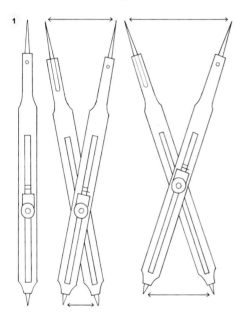

If a microscope with zoom facility is not available, magnifications in multiples of measurements can be made using dividers. **1** Proportional dividers. **2** Enlargements using simple dividers, from left to right showing enlargement – x 1, x 2, x 3.

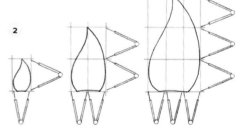

Making a finished illustration

A line drawing for publication should be made on good-quality line drawing paper with permanent light-fast ink, especially if it is destined to become part of a collection of illustrations in an herbarium. The ink must be black so that all details are clear when the drawing is reproduced in print. In order to avoid alterations and corrections which may spoil the smooth paper surface for inking-in, preliminary work should

be done in pencil in a sketchbook as suggested above. It can then be transferred to line drawing paper for the fair copy (see how to do this below).

If there is too much time-consuming detail in a habit sketch for it to be easily traced and transferred, it can be drawn straight onto the good paper. It is still advisable to use a tracing paper layout to make sure there is enough space for all the parts to be included.

Reduction		Scale Bars		Magnification
x 1	2 cms		2 cms	x 1
x ½	4 cms		4 mm	x 10
x ⅓	6 cms		1 mm	x 20
x ¹⁄₅₀₀	1 m		0.5 mm	x 40

Scale bars should be placed close to the object
to which they refer and be long enough to
enable the viewer to estimate sizes easily.

From the sketchbook to the fair copy

After collecting all the components of an
illustration in a sketchbook they can be
traced separately and arranged in position
on a rectangle the size of the illustration
drawn on tracing paper. They can then be
transferred to line drawing paper by the
following method:

On a piece of line drawing paper draw the
same rectangle as that used for the tracing
paper layout. Place the parts in position
on the rectangle. To transfer a tracing, turn
it over, place it on a piece of waste paper
and coat the reverse side with graphite
using a 3B pencil. Wipe off excess graphite
with tissue, turn it back to be right side up
and place it in position. Use masking tape
to lightly fix it in position, then use a hard
4H pencil draw over the lines on the top
side and a faint image will be transferred to
the line drawing paper. Remove the tracing
and neaten the lines of the image using
a sharp 2H or HB pencil, referring to the
original drawing to add any missing details.

Inking in

When the transferred image has been
neatened and excess graphite removed
by dabbing with an eraser, the image
can be completed by drawing over the
pencil lines with a pen and permanent
ink. Complete the outlines before
adding the details. Wait for at least an
hour before erasing pencil lines under
the ink to give it time to dry thoroughly.
Fibre tip pen lines are best left for more
than an hour or overnight to avoid
losing any ink.

Finishing the illustration

The last stage is to make sure the parts
of the illustration are correctly labelled
and remove any unwanted marks. A
signature can be added in a suitable
place, usually somewhere at the bottom
of the illustration.

Corrections

These can be covered using correction fluid or scratched out with a sharp razor blade or scalpel but it may not be possible to re-draw over either of these corrections. Sometimes it is necessary to remove part of an illustration and insert a replacement piece of paper on which a new version can be drawn. To do this use a ruler to outline the area to be removed and cut round it with a sharp scalpel against the ruler (a metal one if possible). On a fresh piece of paper draw the outline of the replacement using the hole as a template and cut it out in the same way. Fit the new piece into the hole and secure it on the back using adhesive tape. A new drawing can then be made and the edges of the insertion removed at the scanning stage.

Captions

The artwork should have the name of the plant illustrated written in pencil on it. Each part can be given a letter, A, B, C etc., written in soft pencil beside it and listed on the back of the artwork with magnifications before and after reduction in printing. This information will be used by the commissioning botanist to write the figure captions for the final publication, so ensure that they are clearly written, legible and logical.

Shading

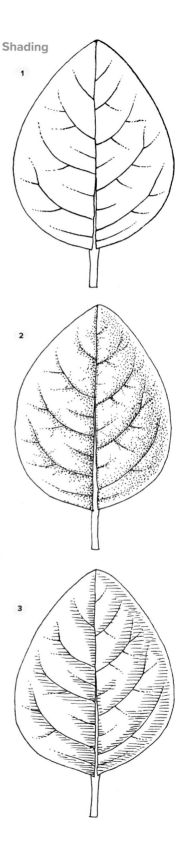

Shading is best kept to a minimum for botanical line drawings from dried specimens but three-dimensional objects, such as fruits and flowers, may be improved by some shading to show form. Indicate the shading in pencil on the fair copy before working over it in ink. Various shading techniques. **1** No shading. **2** Stippled shading. **3** Line shading.

Recommended art materials for line drawing

- Cartridge paper sketchbook
- Pencils, 2H, HB, 3B
- Eraser, good quality, either plastic or putty type
- Sharpener or knife and fine emery paper
- Dividers or compasses
- Proportional dividers
- Tracing paper, lightweight 63 gsm

- Smooth surface line drawing paper such as Bristol board
- Technical drawing pen, 0.18, 0.25 mm nibs
- Black ink for technical pens
- Dip pen and fine nibs such as Gillot 303, Hiro Leonhart H6 or mapping pen
- Black waterproof Indian ink or acrylic ink for use with a dip pen
- Fibre tip pens, good quality 0.3 mm, 0.2 mm, 0.1 mm, 0.05 mm

Art materials for line drawing: *from left to right*: dip pen with Gillott 303 nib and mapping pen with waterproof ink above, technical drawing pen with suitable ink, two makes of fibre tip pens in sizes 0.1 mm and 0.05 mm, traditional wood pencil with sharpener above, 2 mm clutch pencil with rotating sharpener above, putty and plastic erasers, *above*: emery paper mounted on card for maintaining a sharp point on drawing tools, *below*: simple dividers, proportional dividers, 6" ruler.

FLORAL DISSECTION

Christabel King

Floral dissections explore the structures of flowers which are important for their identification and classification. The reproductive phase of plants is considered the most reliable for identifying them from their physical characteristics. These are more constant than vegetative characters, which can be affected by the conditions in which a plant is growing. When studying and naming dried pressed specimens as found in herbaria it is important to be able to examine the internal structure of flowers and reproductive organs. These are usually not possible to see without cutting open a flower and dissecting the parts. Most work must be done from pressed herbarium specimens and it is more difficult to dissect dried material than a living specimen, but dried plant material can be softened and reconstituted owing to the cellulose and woody components in plant cells. Rehydration will usually make it possible to separate the parts and observe the internal structures. Although pressing and drying will have lost the three-dimensional appearance, enough can usually be seen to make a description and/or drawing.

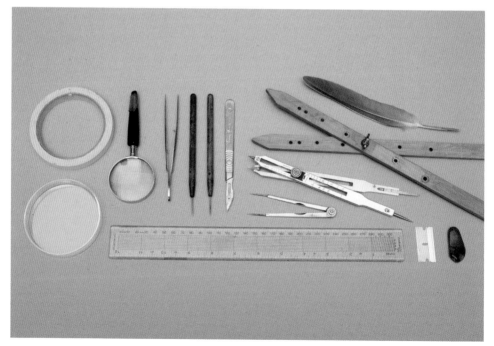

Tools for floral dissections: *Top row left to right*: Adhesive tape, hand magnifier, forceps, mounted needles, scalpel, simple dividers, proportional dividers, large proportional dividers, feather for delicate brushing. *Bottom row left to right*: Petri dish, ruler, razor blade, smooth stone burnisher.

Choice and treatment of specimens

Living material

It is rewarding to study living plant material under the microscope because it reveals flower structures as they are in nature, with the colours and markings which exist as guides for pollinators. Living plant tissues dry up rapidly, especially under a lamp that is not a cold light source, so it is necessary to keep the dissection in a small dish of water while working on it. Afterwards the dissected parts can be preserved for further study as a dried specimen or in preserving fluid. To dry dissected parts, arrange them on a piece of cartridge paper while still in water, then lift the paper with the parts on it so that the water drains away. Excess water can be carefully blotted and the dissection left to dry. It can then be put in a small paper envelope, called a capsule, and kept as a dried specimen.

Dried material

Herbarium specimens are valuable and it is important to be economical with material as most sheets do not have an abundance of flowers. Check whether there is a small paper envelope attached to the herbarium sheet, called a capsule, containing loose flowers, which would make it unnecessary to detach anything from the mounted specimen. Select a flower which is in good condition and not distorted during pressing. Carefully detach it from the sheet using mounted needles, forceps and scissors and place it in a petri dish. Some features such as hairs can be more easily seen on dried material but most parts can only be seen if the material is moistened and reconstituted. This can be done using boiling water or a wetting agent. Ammonia can be used on fleshy flowers which cannot be softened in water. After use, keep the dissected material by drying and pressing it in a small paper envelope (capsule), which can be attached to the herbarium sheet from which it was taken. If there is only one flower, which should not be removed from the herbarium sheet, it may be possible to moisten an area of the corolla and cut open a flap, which can be lifted up so that the internal structures can be seen.

DISSECTION AND ILLUSTRATION OF PLANT MATERIAL
Christabel King

It is not difficult to cut open a flower and observe its structure, but if it is to be drawn for an illustration the dissected flower needs to make sense visually. A drawing of a dissected flower is usually semi-diagrammatic in order to show the parts and structures as they would be if cutting through them did not damage them. It may be necessary to open one or two flowers and observe the structure before making a good enough dissection to draw.

Conventions used when drawing floral dissections

1. Parts are arranged vertically on the page to give an orderly appearance to the layout.
2. Cut edges are shown as double lines, dotted lines or a straight ruled edge.
3. Enlargements of surface textures can be bounded by a rectangle or circle.
4. A cross-section through a part can be linked to that part with a line.

Cutting the flower open: half-flowers and longitudinal sections

A half-flower diagram is useful because it shows the natural position of organs within a flower. It can be used for any type of flower but if a drawing is to be made of a tubular flower, a profile to show its shape should be drawn before cutting as this is likely to be spoiled when the cut is made. The cut should follow the line of the central axis, dividing the flower in half longitudinally. As it is difficult to section small and delicate parts in the centre of the flower, these can be left whole and called a longitudinal section.

Some examples of half-flowers and longitudinal sections

Another way to cut open a zygomorphic tubular flower is to separate the upper and lower parts of the corolla. The two parts can then be placed side by side for comparison. This reveals the shape of corolla lobes better than a half flower.

Cut lines (in red) for ½ flowers. In **1** and **2** there is only one way to cut the flower in half, whether you start from a petal or between a petal. In **3** and **4** there are 2 ways to cut it in half either through 2 petals or between 2 petals. In **5** there is only one way to cut it in half.

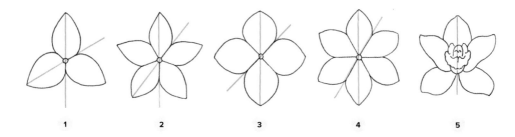

1 2 3 4 5

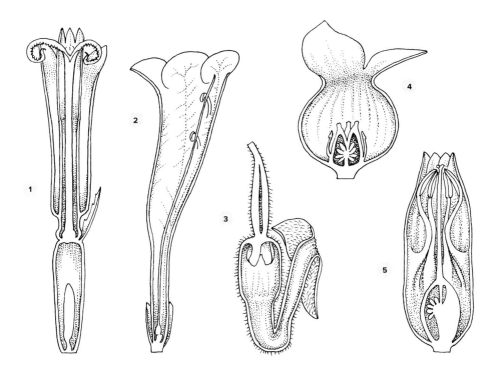

1 *Helianthus salicifolius.* **2** *Strobilanthes hamiltoniana.* **3** *Aristolochia* sp.. **4** *Asarum sieboldii.*
5 *Polygonatum kingianum.*

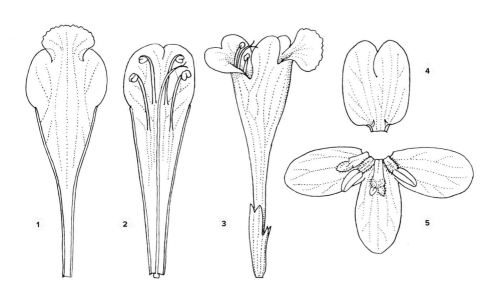

Meehania fargesii. **1** Lower half of corolla. **2** Upper half of corolla. **3** Whole flower.
Petrocosmea thermopuncta. **4** Upper portion of corolla. **5** Lower portion of corolla.

FLORAL DISSECTIONS
Christabel King

Actinomorphic flowers

The parts of the flower can be removed from the peduncle in order, from the outermost to the centre and then arranged in a line. If there are only a few parts all can be measured, otherwise a representative range can be taken, for example the largest and the smallest with one of middle size.

Leontice leontopetalum
1 Stamen, side and abaxial view
2 Whole flower
3 Petals
4 Gynoecium with part removed to show ovules
5 Nectary, 2 views
6 Apex of style

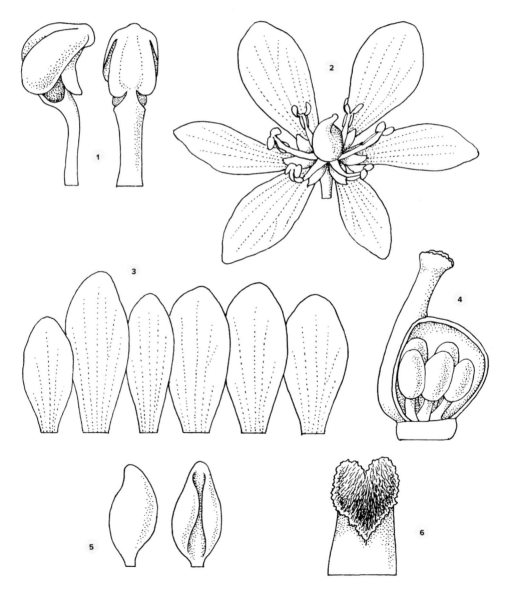

Asymmetrical flowers

For flowers without any noticeable regular symmetry, start the dissection from the point of attachment. The parts can be removed one by one from the outermost to the centre, in the same way as for actinomorphic flowers.

Lathyrus latifolius
1 Flower, front view
2 Whole flower side view showing parts to be removed from outer to inner
3 Dorsal petal
4 Calyx
5 Petal
6 Petal
7 Keel
8 Stamens
9 Gynoecium

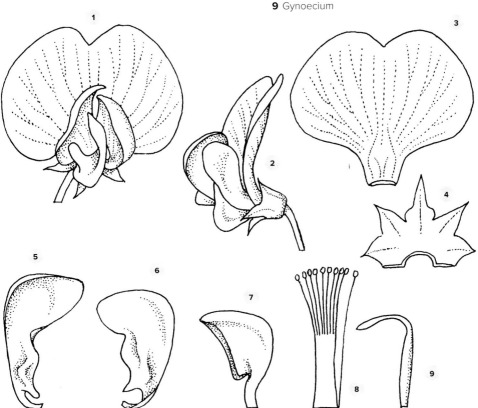

Measurement

Botanical descriptions include measurements of the parts of flowers. Parts large enough can be measured with a ruler but for small parts a graticule in the eyepiece of a microscope is used. This is a scale on a piece of glass inserted in the eyepiece of a microscope with divisions of 1/10 mm. When the microscope zoom is set on x 1 it can be used to measure the size of an object directly. If another setting is used, for example x 2 or x 4, the size must be calculated by dividing the measurement by 2 or 4. The graticule can be checked for accuracy by putting a ruler with millimetre divisions under the microscope and checking if the divisions of 10 mm coincide.

IDENTIFICATION AND NAMING
PRIORITISING SPECIMENS FOR IDENTIFICATION
Martin Cheek

Identification of specimens is a major function for many herbaria. While DNA barcoding and artificial intelligence offer great potential for plant identification, to date, identifications depend on speedy physical access to authoritatively named herbarium reference specimens by competent taxonomists.

Low and high priority specimens

Specimens arriving for identification can be generally divided into two streams, low and high priority:

Low priority specimens may not require naming urgently; examples might comprise a batch of specimens from a fieldwork project, perhaps ancillary to a doctoral thesis. They can be named to family and genus and then be accessioned and incorporated in the herbarium within the genus for the geographic area concerned. They will then be available for visiting specialists, or when time permits, taxonomists of the herbarium concerned, to identify.

High priority specimens require naming urgently; examples may include a plant-poisoning investigation, or an environmental survey of an area being considered for development where it is important to know if any species are threatened with extinction and the conservation value of the area. These specimens should be named to species using the following workflow.

Identification to species

- Specimens are first sorted into families, then into genera within those families based on morphological characteristics and field data. If available, specialists should be consulted to perform identifications at least to genus level, but many plant groups lack such specialists

- Identification keys, such as Floras, field guides and taxonomic revisions and monographs, if available, are then used to place tentatively each unidentified specimen with a species within the geographical area concerned

- In some groups, parts of the specimen such as the flowers or fruits, will need to be rehydrated to examine characters important for species placement

- The tentatively identified specimen is then physically matched by eye with authoritatively identified specimens of the species concerned, using a microscope where needed, to confirm matches with less tangible characteristics such as venation and indumentum details of which are difficult to convey in words or in a key but which the human eye and brain can detect

- As an insurance, material of neighbouring, similar species in the herbarium arrangement should be consulted to check that there is no mistake: identification keys do not always provide perfect results

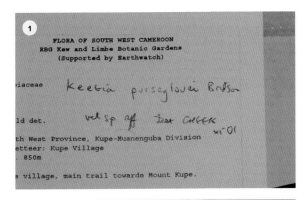

1 Provisional determination of a specimen; note the reference '*vel* sp. *aff.*' = 'or related species'.
2 Provisional determination of an undescribed species; the botanist has indicated the specimen is a new species and a suggested related species. **3** A determination showing that initial identification was not definitive; pencil notes on the specimen show a provisional working name 'sp. *B*.'.
4 Determination slip showing name of the botanist and date of determination; this specimen has been determined as the type a new species and shows the new epithet chosen by the botanist.

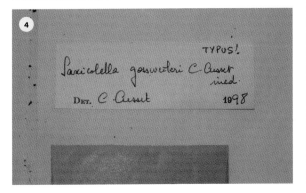

IDENTIFYING NEW SPECIES
Martin Cheek

About 2000 new species to science of vascular plant are published each year, mainly from tropical areas where many biodiverse countries are incompletely surveyed for their plant species (Cheek *et al.* 2020). The vast majority of new plant species to science are detected using morphological characters only, unsupported by other lines of evidence, and good herbarium specimens are fundamental to identifying and describing new species.

Species not matched in the herbarium

If the unidentified specimen does not fall within the range of morphological variation of a species (as represented by reference specimens in the herbarium) within the geographic area concerned, neighbouring geographical areas should be included, since the specimen may represent a range extension of species. If this fails to provide a match, then there are two possibilities:

a) it may be a species not represented in the herbarium (in which case other herbaria, including those online with digital images or via contact with the curator, should be checked for any missing species, accepting then that matching from an image may be uncertain since microscopic confirmation is not possible);

b) the generic or even family assignment may be incorrect: this should be double-checked. If the generic placement is confirmed, and no match to species has been made, then it is possible that the specimen may represent an invasive species, possibly from another continent, or a new species (or other taxon) to science.

Confirming a new species

If a new species is indicated, the first step is to determine to which species or group of species it appears most closely similar (and is possibly related

to). A taxonomic revision of a genus will help greatly with this step.

In formally describing a new species to science the International Code for Nomenclature requires a diagnostic statement indicating the characters by which the postulated new species differs from that or those most similar to it.

It is normal to annotate determination slips for the specimens concerned "sp. nov.", or more helpfully "sp. nov. aff. [give name of most similar species]". In contrast "sp. aff. [give name of most similar species]" indicates that the specimen has not been matched but is considered to be a species close to that named, suggesting that it has not been definitively concluded that the specimen is a new species to science and implying that all the species options for matching may not have been exhausted. The annotation "cf. [name of most similar species]" suggests that the specimen be compared with material of the named species, indicating that it either may be within the range of variation of that species, but that that is uncertain, and that it may represent a related species, including one that is new to science.

A selection of species new to science: all will have type specimens deposited in herbaria. **1** An extremely rare species, *Pseudohydrosme ebo* (Araceae), is restricted to the Ebo forest of Cameroon (Cheek *et al.* 2021). **2** *Barleria thunbergiiflora* (Acanthaceae), was named new to science in 2021 after being collected on a National Geographic expedition to Angola (Darbyshire *et al.* 2021). **3** *Ardisia pyrotechnica* (Primulaceae) was found in the forest in Borneo and named for its resemblance in flower to an exploding firework (Julius *et al.* 2021). **4** *Philibertia woodii* (Apocynaceae), is a periwinkle from the Andean Valleys in Bolivia (Keller & Goyder 2021).

Grouping unnamed specimens

It can happen that a specimen being identified matches no named species, but that a match, or matches, are found in the material named only to genus and awaiting identification to species in the folders at the end of sequence of species of the genus. In this case it is useful and normal to annotate the specimen "= (collector name and number of the matching specimen(s))". Linking specimens together in this way, which are likely to represent a new species to science, is helpful for future taxonomists who may later revise the genus and eventually formally name the taxon to which these specimens belong.

> **TIP**
> The steps that start from being unable to match a specimen, to formal publication of a new species to science are resource demanding. Since resources are often limited, it may be many years before a specimen considered to be a putative new species is published as such. A further source of delay was the former convention by which it was considered poor practice to name a new species based on only a single specimen, based on the possibility that this might be a monstrosity or mutant. Until recently in such cases specimens might be annotated "sp. nov., additional material required". However best practice now is to publish such species as soon as possible if it is considered that they might be at risk of extinction. Once a species has a formally published scientific name, registering an IUCN (2012) extinction risk assessment is greatly facilitated, and if a species is on iucnredlist.org the possibility of protecting it from extinction is enhanced.

CREATING LABELS

Yvette Harvey

Labels attached to specimens are a crucial part of the specimen for the user and curator. The minimum expectation is that each specimen has a unique collection label containing its provenance and preferably brimming with useful information about the plant and the habitat in which it was residing.

Other labels attached to the specimen can include slips with name changes, citations, notification of extractions taken for other biological disciplines, notification of ancillary collections linked to the specimens, conditions governing use, barcodes and detachable loan numbers.

Information to include on specimen labels

It is wise to remember that the label will have an international audience so use simple terminology where possible and include the following for a basic label:

- Your institution (in full, latinised or official abbreviation)
- Family
- Name and author (preferably with standard abbreviation)
- Location (including country)
- Longitude, latitude and altitude
- Habitat [and frequency within the location can be useful for IUCN assessment]
- Habit (ideally in the order in which a plant description is written for reader ease)
- Other notes (i.e., local names (and the local groups/area responsible for its use), ancillary collections (fruits, wood, ethnobotanical notes [within the confines of Access and Benefit Sharing legislation], photographs, DNA, alcohol)
- Duplicates (as herbarium codes)
- Collectors (all collectors present, do not abbreviate additional collectors to *et al.*)
- Collection number
- Date (month in roman numerals to avoid mix up)
- Restrictions

Additional considerations

The specimen may be used differently in 100 years and your audience cannot see the canopy or angle of big branches, the proximity to other plants, their spread, or even inhale its intoxicating fragrance so it is worth noting these and other additional considerations, including:

- Documentation (i.e., permit reference numbers)
- Barcode number (handy for duplicate cross-referencing)
- Notes for your own project/collection (if relevant)
- Go outside your field/the purpose of collecting (e.g., even if flower colour is the only note you need, include other habit details to make the specimen of more use to later researchers)

Try not to over-describe, use too large a font, place information on the reverse of the label, include maps or anything that adds to the size of your label. Not only will they take up too much space on the specimen but may also need folding or gluing in a special way that will make it harder to curate or digitise.

The label should last as long as the specimen, so use archival paper and inks in their production. Remember that mounting is likely to involve liquids, so select a permanent ink that is not water-reactive.

Consider how to attach the label:

- Non-tear paperclips that won't corrode
- Glues that are archival and non-animal based (reduce possibility of pest damage)
- Pins that won't corrode

FURTHER READING

Royal Botanic Garden Edinburgh (2017); Royal Horticultural Society (2015); Victor *et al.* (2004).

HERB. HORT.WISLEY (WSY)
Nemesia RHS Trial 2021 #2124

Nemesia 'Easter Bonnet' PBR

SCROPHULARIACEAE

Trial entry no.: 88
Locality: Trials field, RHS Garden Wisley, Surrey, UK.
Description: Herb to 30cm; stems Strong Yellow Green 144A; leaf upper surface Moderate Olive Green 137B, lower surface Greyish Yellow Green 148C; corolla spur Pale Greenish Yellow 160C, upper lobes Pale Purple N75D, towards throat bluer veining Light Purple 85A, lower lip Light Yellow Green 2C, palate Brilliant Yellow 7A.
Source: Earley Ornamentals Ltd, York Road, Thirsk, North Yorkshire, YO7 3AA
Coll.: Liz Munson & Yvette Harvey
Dup.: A
Date: 6.ix.2021
RHS barcode: WSY0167040

1 Regrettably little can be gained from reading this label, written some years after Sibthorp's death. Our aim today is to give the researcher the ability to interpret the specimen in the most useful way. **2** *Psychotria samoritourei.* Provenance is at the heart of a modern label. Among other pertinent information, all collectors are acknowledged, the specimen has a unique number, and the precise location is geo-referenced. **3** *Nemesia* 'Easter Bonnet' PBR. Note that for a specialist collection like the RHS's herbarium, precise colour is very important and the plant collections' standard for this is the *RHS Colour Chart* (2015: sixth edition).

THE HERBARIUM KNOT
Clare Drinkell

A knot is never 'nearly right': it is either exactly right or it is hopelessly wrong, one or the other; there is nothing in between (Ashley 1944). Herbarium specimens, bundled in corrugated boards and tied using an herbarium knot, can be safely carried within an herbarium but also packaged up to send out of an herbarium.

Herbarium knot
String and cardboard can be used as a simple but effective way to bundle and carry mounted and unmounted herbarium specimens. The herbarium knot is used for this purpose — it is a secure knot that does not slip under tension and can be easily untied to release.

Equipment
String, ideally nylon, approx. 3.5–4 m in length, with a fixed loop knot (known as an overhand loop) tied at one end.

Two herbarium-sized cardboards with corrugates running lengthways.

Specimens needing to be transported (mounted or unmounted).

Tying the knot (for right-handed people)
1. Set the bundled specimens in front of you, on a table in portrait position, with two cardboards sandwiching the specimens.
2. Take the string with the ready-made overhand loop knot in your one hand and thread it around the width of the bundle, passing the opposite end of the string through the loop knot and tightening the join on the centre top of the bundle.
3. Holding the join in this position with the string tight around the bundle (adjust to tighten if necessary), pass the string over the top end of the bundle, underneath, and back up to meet the join, making a cross.
4. Thread the end of the string around the back of the top arm of the cross in a left-to-right direction and pull tight. Adjust the join if it has gone off centre of the bundle.
5. With 10 cm of the remaining string held in your left-hand thumb and forefinger, make a 'D' shape in the bottom right quadrant, clasping together with the bottom leg of the cross. With right-hand thumb and forefinger, reach through the 'D' hole to pull through a hoop handle from the remaining string.
6. Adjust and tighten the string, keeping hold of the hoop handle.
7. To release, pull the end.

> **TIP**
> Do not be tempted to turn the bundle upside down when tying up a pile of specimens.

HOW LONG IS A PIECE OF STRING?
The ideal length should measure approximately 3.5–4 m long including the loop for a bundle approximately 15–20 cm in height when tied, but this is dependent on the size of your herbarium sheets.

The 7 steps to tying an Herbarium knot

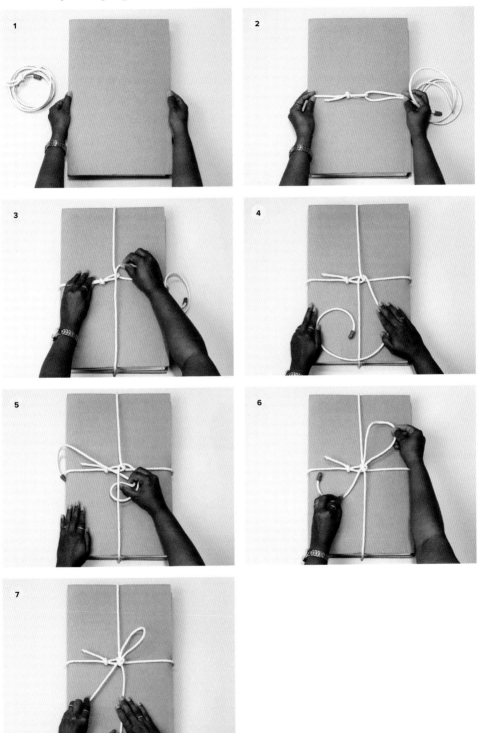

PROCESSING UNMOUNTED SPECIMENS
MATERIALS
Lesley Walsingham and Renata Borosova

To prepare plant specimens for mounting, it is best to have a range of equipment available to use on different types of collections, for example secateurs for removing excess woody parts. Capsules should be prepared at this stage to be mounted alongside the specimen.

Materials used for curating and preparing specimens before mounting

Equipment:

- Spare newspapers/flimsies: to enclose unmounted specimens
- String: to secure bundles for transit and storage within the herbarium

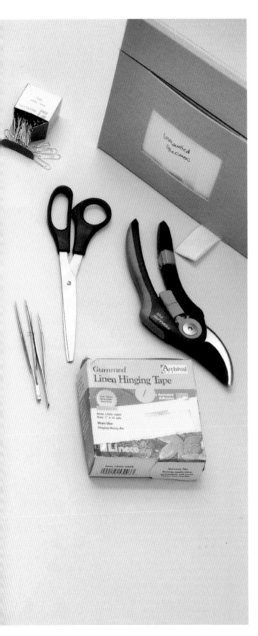

- Cardboard supports: to be placed at the top and bottom of bundles of unmounted and mounted specimens to support and protect them. The cardboard should be no larger than the mounting sheet
- Capsules: glued to sheets during the mounting process to contain and protect small portions of a specimen, or very small specimens or entire specimens such as Bryophytes. Capsules can be made by folding a rectangular piece of paper (ideally archival quality)
- Pens (ideally archival). At Kew, an archival red pen is used to note on the label when there is associated material such as carpological/bulky specimens or when there is more than one sheet
- Tweezers: for removing delicate plant parts
- Pencil: medium soft pencils (HB, 2B or F) are ideal for writing on capsules the collector and collector number and can easily be erased/corrected
- Paperclips: to ensure the capsules are securely closed. Continual-loop plastic clips are best for humid tropical conditions. Brass paperclips do not rust or stain the paper. 'No-tear' metal paper clips are acceptable for drier conditions where they are unlikely to become rusty
- Translucent glassine envelope: to contain seeds inside a paper capsule
- Secateurs, scissors and scalpels for removing excess material
- Linen tape: for ferns
- Rubber
- Sharpener

Materials used for curating and preparing specimens before mounting.

PREPARATION OF HERBARIUM SPECIMENS BEFORE MOUNTING
Renata Borosova and Lesley Walsingham

Preparation of plant specimens before mounting ensures the mounting process itself happens smoothly and efficiently. Important checks verifying material and suggesting its eventual display for giving maximum observation are extremely useful at this stage. Herbaria may carry out preparation checks as a separate task; smaller herbaria may combine them with the mounting process itself.

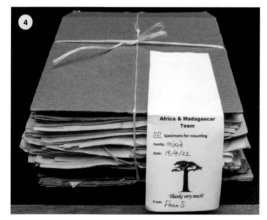

1 Check collector's number on the jeweller's tag and on the label match, so mixed collections are not mounted. **2** Specimen prepared for two separate sheets. Identical labels with each, one label annotated 'Sheet 1 of 2' the other 'Sheet 2 of 2'. **3** Important loose parts in secure capsule. **4** A secure, manageable, well-labelled bundle, for transfer to mounting. **5** Specimen leaf trimmed to fit sheet, but important characteristics of length, range and shape still displayed on the other leaves.

Why prepare specimens before mounting? Aims and checks

The main objective is to select appropriate material for the best presentation layout possible prior to mounting and preserving the specimen onto the sheet, ensuring the maximum amount of information is available for future studies.

AIMS

- Select the most useful plant material to be mounted
- Choose plant parts to remain loose in capsules, for detailed study and DNA extraction
- Put aside excess material suitable to be sent as duplicates (see page 112)
- Organise material for ease and speed of mounting

CHECK

- Label is present, preferably on archival paper
- Specimen is named to family and genus level on the label or determination slip, whenever possible
- For pest damage, report and take further action if necessary (see: Pest management)
- Compliance with institution policy on CITES and Access to Genetic Resources and Benefit Sharing agreements and legislation (see: Acquisitions and disposal)

Preparing process

- Ensure each specimen fits a mounting sheet; some may require 2 or more sheets. Fold or trim where necessary, leaving room for capsules and labels
- Remove as much soil as possible and dispose of this and excess material in a biosecure way
- Separate plants in clumps (unless this shows the growing habit at its best)
- Place in an appropriate capsule: small loose delicate parts, important parts trimmed from the plant, excess flowers

and fruits and loose—leaf pieces - suitable parts for DNA extraction and anatomical work in the future. Any photos that accompany specimens should also be placed in the capsule for the sheet, to preserve the image with the collection

- In some cases, a glassine envelope may be used to contain delicate flowers or tiny fruits/seeds. These envelopes are then placed within the capsule
- If pre-cut capsules are not available, use a folded piece of archival quality paper. For guidance on how to fold capsules see Victor *et al.* (2004). Make sure small parts cannot escape from the corners
- Display both sides of leaves and petioles, leaf base and tips. If there is only one leaf, cut a section and mount or keep loose in a capsule
- Show a full range of leaf sizes if possible
- If the specimen has very bulky parts, e.g., fruits, carefully remove these and place into a box with a copy of the specimen label. Ensure that this is cross-referenced to the herbarium sheet (see: Processing bulky collections)
- Expose any hidden flowers and fruits, removing leaves if necessary
- Display the front, back and sides of flowers where possible
- Trim projecting spines or branches to prevent damage of other sheets in the herbarium cupboard
- For specimens to be mounted on two or more sheets, photocopy the label for each sheet and clearly mark (Sheet 1 of 2, 2 of 2, etc.)

> **TIP**
> Use sharp-pointed tweezers for moving fragile material or slip a piece of stiff paper under material to move it around.

PROCESSING DUPLICATE MATERIAL FROM SPECIMEN COLLECTIONS SENT TO HERBARIA

Renata Borosova

Duplicates provide extra material that can be stored in other herbaria to expand their collection and to ensure it is available elsewhere should anything happen to the original material.

Aims and checks

To distribute extra material to other herbaria as duplicates to enhance their collection and further their research purposes.

CHECK

- The terms and conditions of collection and acquisition allow the transfer of duplicates
- Duplicates can only be sent to third parties with prior consent of the provider country
- All restrictions on the use of specimens covered by the collecting permit and material transfer agreement must be observed
- There is sufficient decent quality material to separate and send as a duplicate to another institution
- Material selected shows a good range of useful characteristics
- If a unisexual plant, both female and male flowers are represented
- The range of leaf sizes and shapes is represented
- Flowers and fruits are included, wherever possible
- Parasitic plant and its host are kept together

Process and distribution

- Separate selected duplicate specimens, each into a separate flimsy or newspaper
- Add photocopied labels and determinations
- Add an indication of distributing herbarium
- Select herbaria with appropriate family or geographical interest, using compiled distribution lists if available and again checking against the original terms and conditions relevant to the collection
- Prepare secure, manageable, well-labelled bundles for transfer to distribution process

> **TIP**
> Best practice is not to send out sterile material, unnamed or dubiously named specimens. Parts of fern fronds should not be sent as duplicates.

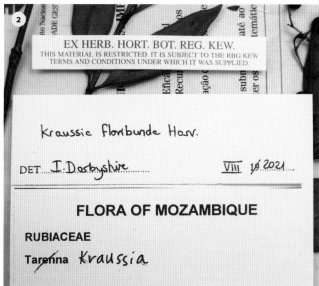

1 Kew specimen and duplicate specimen prepared for distribution.
2 Addition of an 'Ex Herb. Hort. Bot. Reg. Kew.' label to indicate the distributing herbarium and conditions of distribution.

PLANT FAMILY GROUPS REQUIRING SPECIAL PROCESSING

Lesley Walsingham

Plant families featured here have characteristics that require additional or specific care and attention.

Ferns

- Only fertile, spore-bearing material should be mounted; discard unfertile material in a biosecure way
- Many sheets may be needed for one collection
- Sheets should display a complete frond
- When separating fronds to fit sheets, use marked tape to indicate where parts were attached
- Sheets should be numbered in sequence so the complete picture of the frond can be reassembled if needed
- Ensure the specimen prepared displays scales (on rhizome, base of stipe, or elsewhere)
- Spores may need to be put in a glassine packet, then a capsule

Palms, Cacti and Gymnosperms – dealing with bulky specimens

- Strong folders with added depth will accommodate some specimens
- Some larger parts such as stems may be attached to extra firm boards, usually by stitching
- More bulky specimens are best in large archival top-fitting lidded boxes, which are different to carpological boxes and allow all parts along with collector's data label, to be kept together
- Cactaceae flowers, fruits and whole stems ideally preserved in spirit, can also be dried and cut to fit carpological boxes
- Ideally some specialist training of cutting specimens to fit is given to curators
- Boxes and folders must be correctly and clearly labelled inside and outside with duplicate labels

Other groups

Compositae

- Should have whole capitula placed in capsule where possible

Poaceae and Cyperaceae

- Sterile material should not be mounted
- Inflorescence, spikelets, fruits, transition leaf sheath-lead blade are clearly displayed
- Some spikelets loose in a capsule are useful for dissection

Tiny plants

- If numerous, allow some for attaching to sheet and put some in a capsule; if few plants, put all in capsule for attaching to sheet

Algae and Bryophytes: floating or submerged aquatics

- Float specimen out onto mounting paper on wire mesh and in tray of water, for best display
- Dry specimens quickly but not with excessive heat
- May have to remain on drying paper, cutting around to mount on sheet

Succulent or fleshy plants

- May require slicing or halving to reduce bulk and are best preserved in a spirit collection

> **TIP**
> Be aware of plants that may have irritant hairs, e.g. *Mucuna* (Leguminosae) or the stinging hairs in some Urticaceae. Ensure PPE is available to wear as appropriate.

1 Fern marked up with linen tape to indicate attached parts. **2** Fern from one collection showing top-to-bottom display. **3** Palm specimen in large archival top-fitting lidded box. **4** Palm specimen in strong folder with added depth. **5** Grass specimen, leaf sheaths clearly displayed, spikelets loose in capsule. **6** Whole Compositae capitula placed in capsule. **7** Gymnosperm, less bulky specimen stitched to extra firm boards. **8** Cactaceae in carpological box. **9** Tiny plants some on sheet, some in capsules.

PROCESSING SPIRIT MATERIAL

Melissa Bavington

Material to be fluid preserved should be transferred to Kew Mix as soon as possible after collection. This work should be carried out in a fume cupboard. When material is ready to be accessioned into the collection it should then be transferred to collection jars.

Processing spirit

For long-term storage of botanical specimens in spirit, the specimens should be fixed and preserved; there are many variations of fixation and preservation mixtures.

Kew Mix contains

5% formaldehyde (fixative)

5% glycerol (to prevent plants from becoming brittle)

37% water (to achieve desirable dilution)

53% industrial methylated spirits (preservative)

Jars should be appropriately sized to adequately fit the specimen and label – if the specimen or label is above the fluid level it will act as a wick and the fluid can evaporate faster. Place the label at the base of the jar and specimen on top so it is not impeded by the label. Simmons recommends a fluid-to-specimen ratio of 7:3 (Simmons 2014).

Specimens that arrive via donations or collecting trips in a temporary fluid (ideally labelled to aid specimen processing into the collection store) should be rinsed thoroughly to remove the original fluid. It can then be placed in an alcohol bath where the concentration of alcohol is stepped up. The specimen should be monitored; when the concentration of alcohol reaches 70%, it can then be placed into storage as is or into Kew Mix.

If there are smaller parts or a dissection is made from the specimen, it should be placed inside a smaller-sized tube with a push-in closure that can fit inside the original jar.

When specimens are required by researchers, they are transferred to a 'safer' mix, namely one that does not contain formaldehyde.

Copenhagen mix contains

70% industrial methylated spirits

28% water

2% glycerol

Storage

For the most efficient use of space, it is recommended that spirit material is kept in specimen/bottle number order rather than taxonomically, and the collection recorded on a database. The jars should be kept in metal drawers so they are shielded from light, ultraviolet light and vibrations (Simmons 2014). Where possible bespoke metal drawers to fit the jars are recommended. The overall environment of the collection store should be cool and stable; higher temperatures accelerate deterioration and evaporation, while lower temperatures cause preservative problems (Simmons 2014).

Other considerations depend upon how different solutions withstand different temperatures, the balance of fire precautions measures and human comfort levels. Specimens in Kew Mix store well at 12–14°C. Relative humidity should be at 50% RH.

TIP

All work with spirit specimens should be carried out with the appropriate personal protective equipment (gloves, lab coat, respirator if appropriate). A fume cupboard should be used when opening jars of Kew Mix.

FURTHER READING
Neumann *et al.* (2022)

1 Topping up fluid in a storage jar. **2** Filled drawers of various sized storage jars. **3** Pre-numbered storage jar lids. **4** A fluid-to-specimen ratio of 7:3 with label and specimen fully submerged by the fluid (ideal specimen storage).

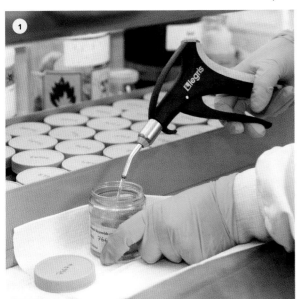

PROCESSING ETHNOBOTANICAL COLLECTIONS
Ben Hill and Mark Nesbitt

Ethnobotanical specimens vary greatly in shape, size and format. Careful use of different storage formats will protect specimens from environmental damage and incorrect handling (Balick & Herrera 2014; Timbrook 2014). Special considerations for ethnobotanical material include safe handling of poisons and preservation of specimen chemistry and DNA.

Glass jars

Loose material, such as seeds, powders, and liquids should be stored in airtight glass containers with plastic seals or lids, with labels affixed to the exterior. Historical glass containers, sealed with cork or glass stoppers, should not be replaced unless necessary, as both packaging and labelling can be highly informative. If a change of container is essential, labels should be transferred to the new jar and digitally recorded (if possible).

PROS
- Greatly reduces the risk of pest or mould outbreak due to the airtight environment
- Greatly reduces the risk of contamination from other parts of the collection
- Airtight conditions help preserve chemistry and DNA

CONS
- Potential to smash if handled incorrectly
- Are often space-intensive as they cannot be stacked

Boxes

Boxes used to store objects must be made from acid-free board or an inert corrugated plastic such as correx. They need enough room inside to allow safe retrieval of objects. Objects must be clearly visible within a box and should not be stacked on one another inside.

A small photograph of the specimen on the box exterior is useful. Boxes containing poisons or sharps, or both, should bear a hazard warning and handling instructions (Banks 2015).

PROS
- Space-efficient as boxes are able to be stacked on top of one another
- Easily moved between workspaces

CONS
- Less protection of contents from environmental conditions than with glass jars
- Pests can move from one box to another

Packing material

Packing materials should be used to cushion and support objects within a box, reducing vibration and friction. Acid-free paper is the most cost-efficient packing material. Sheets can be folded in on one another to create pillows for support and to fill excess space. Paper should be replaced every few years as it will absorb contaminants from off-gassing by the specimen and turn a shade of yellow. Polyethylene foams such as Plastazote® and Jiffy sheets come in a variety of thicknesses and densities. They are most often used to store highly fragile objects, as they can be cut into custom-made supports. Other commonly used materials include inert polyethylene sheeting (Tyvek®), cotton bolsters and linen tape. Any products containing plasticisers should be avoided.

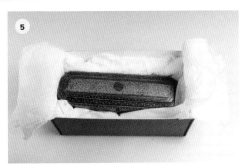

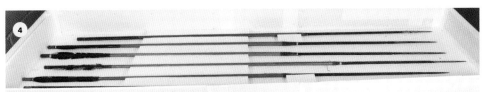

1 Loose material within glass jar. **2** Assorted storage glass jars. **3** Examples of box types and sizes. **4** Example of poison arrows in storage. **5** Standard packed box. **6** Acid-free sheets and pillows. **7** Acid-free tissue discoloration. **8** Plastazote packed box.

PROCESSING BULKY COLLECTIONS
Xander van der Burgt

Dried specimens that are too large to mount on herbarium sheets are stored separately as bulky or carpological collections. They may be used for taxonomic studies, artworks and museum displays.

Curating a bulky collection
The bulky specimen and label are placed in a box. A glass lid will keep the contents of the box visible. Boxes of different sizes should be available; the smallest box that fits the bulky material must be selected. A small label with the country, scientific name, collector and number should be attached to the lid. Alternatively, bulky specimens can be placed inside archival quality paper bags, and the label and barcode are glued to the outside of the bag.

Storing a bulky collection
Boxes or bags with bulky collections are conveniently stored on trays, on shelves, or in shallow drawers. Bulky collections should be stored separately from the herbarium sheets, but ideally kept as close to them as practical. The sequence of families and genera should be the same as in the herbarium, but for simplicity the species can be arranged alphabetically. Very large collections that do not fit in boxes can be kept on shelves in closed cupboards.

Cross-referencing a bulky collection
All bulky material should be databased, barcoded, and imaged. The specimen database should provide the cross-reference with the corresponding herbarium sheet. A cross-reference can also be written or printed on both labels, in red, archival quality ink. If a collection consists of bulky material only, an herbarium sheet bearing only the label and cross-reference may be added to the herbarium.

1 A storage box for bulky collections, containing seven collections, each in a separate box with a glass lid. **2** A storage box with a glass lid, containing a single collection and its label, consisting of an infructescence, fruits, and seeds.

PROCESSING SEEDBANK VOUCHERS

Daniel Cahen

Seedbank vouchers are processed in much the same way as any other herbarium specimen but there are specificities to consider.

Identification

Seedbank vouchers are often crucial to identify seed collections (ENSCONET 2009). Any new identification of a voucher specimen should be added to a database that links it to its associated seed collection.

Cultivated vouchers

If no wild voucher was collected or if it cannot be identified, it is sometimes possible to regrow a cultivated voucher from the seeds. Whether a voucher is wild or cultivated should be recorded on the specimen label and in the seed collection database. Cultivated vouchers may differ morphologically from wild individuals or misrepresent the seed collection if seeds of different species were accidentally mixed when collected.

Restrictions

Seedbank vouchers are incorporated into the general herbarium collection because they are also standalone herbarium specimens in their own right. However, specific restrictions may apply depending on what agreements are in place with the institution that collected the seeds. It may not be possible, for example, to loan, sample, image material or share label data. This should appear on the specimen label to inform users of the herbarium.

FURTHER READING
ENSCONET (2009).

1

> **MILLENNIUM SEED BANK PROJECT**
> **RESTRICTED PLANT MATERIAL. NOT TO BE LOANED, SAMPLED, COPIED OR DIGITISED IN ANY WAY BY NON-KEW STAFF WITHOUT WRITTEN PERMISSION.**
> Loan/sampling requests and new determinations to be sent to the Millennium Seed Bank Project Herbarium Liaison Officer, Herbarium, RBG Kew, Richmond, Surrey, TW9 3AE, UK. Email: msbhlo@kew.org

1 Seed bank voucher label indicating possible restrictions on the material. **2** Cultivated vouchers can sometimes be regrown from seeds.

MOUNTING HERBARIUM SPECIMENS
MATERIALS
Martin Xanthos

The following list comprises the equipment and materials used to mount herbarium specimens. All items listed should be of archival quality.

Materials for mounting specimens

Equipment:

- Mounting paper or white board – for bulky specimens. It is useful to have the herbarium name printed on the mount

- Archival adhesive or other suitable glue – acid-free archival adhesive, water soluble, reversible, does not colour over time and will not damage the surface or specimen

- Plastic squeeze bottle – for applying adhesive to delicate plant material

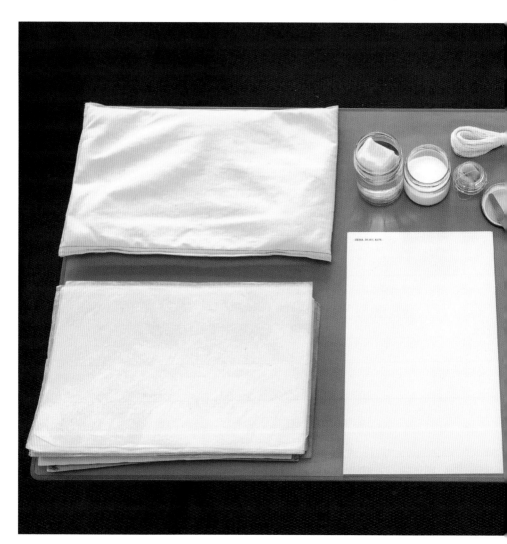

- Brushes, spatula – various sizes for applying adhesive more broadly, e.g., to the labels
- Sponge – to absorb any excessive glue using water to dilute
- Waxed paper sheets – Mount sized sheets, used to prevent excess adhesive sticking to blotting paper while drying and under applied pressure
- Blotting paper – Mount sized papers, used to absorb moisture from the adhesive and aid drying

- Sandbags – Mount sized, heavy-weight canvas bags filled with sand. Used to apply pressure during the drying process
- Wadding – Mount sized, used during the drying process to avoid causing pressure damage to the specimens
- Small weights – to apply pressure to parts of specimen or labels while glue sets
- Capsules of different sizes – to hold loose plant material or labels
- Glassine paper – 'windowing technique' for mounting delicate plant parts
- Non-tear paperclips – to securely close the capsule
- Forceps – for handling delicate specimens, especially useful for moving the glued specimen to the mount position
- Adhesive tape – used for the strapping method
- Nylon or thread and needle – for the stitching method
- Gummed paper – used to cover the knot in the stitching method
- Secateurs, scissors – for trimming oversized specimens

Table arrangement with equipment for mounting specimens.

GENERAL PROCEDURES AND PRINCIPLES
Martin Xanthos

Once the unmounted specimen, associated labels, etc. have been prepared, the next stage is to attach the specimen to archival support. The aim of the mounting process is to provide long-term storage, with the displayed specimen allowing for maximum observation so that all its features can be examined scientifically (Carter & Walker 1999).

Organisation is key

Some herbaria have a two-step process for mounting specimens:

1. Preparing the specimen for mounting
2. The mounting process itself

Smaller herbaria will tend to merge the two-step process. General principles here outline the stage 2 process once specimens have already been prepared (see: Processing Unmounted Specimens). The stage of attachment of the specimen to the mount should be a reversible process, in case the plant material needs to be detached at a later point.

Evaluate the prepared specimen

Have a clear space in which to work. Arrange specimens and equipment for the mounting process so that everything is close to hand. Each prepared specimen needs to have the same preliminary checks prior to attaching to the mount:

- Check that the plant material matches the label
- Ensure that the specimen will fit on the mount, allowing for border space
- Ensure that all parts of the specimen can be clearly observed once mounted

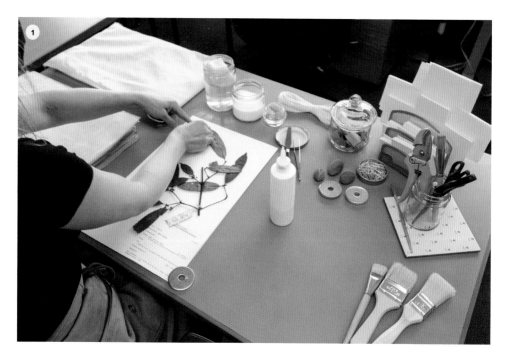

- Do the specimen, labels, capsules, etc. fit on a single mount? If multiple sheets are required, have the labels been prepared accordingly?
- Is part of the specimen separated in an auxiliary collection, e.g., spirit or carpological material, and if so, do the labels indicate this?
- Observe the specimen and establish the most appropriate mounting technique
- Consider how bulky the specimen is and whether it requires a heavier mounting card
- Do not discard any of the prepared specimens, capsules, label(s), etc.

Adhesive is an important component of the mounting process. Minimal use of glue is preferred when attaching plant material to the mount to allow for subsequent removal of the specimen, if necessary. The choice of adhesive used is wide-ranging from natural (animal-based) to synthetic. Croat (1978) and Clark (1986) discussed the merits of both water-soluble and plastic-based adhesives while Grenda-Kurmanow (2021) conducted a survey on the different adhesives preferred by institutions around the world and found the choice was dependent mainly on the adhesive's working properties, long-term chemical stability and compatibility with the plant material.

TIPS

Several plant species contain irritant/stinging hairs or latex, which can be a skin irritant. Consider wearing gloves when handling such specimens.

If specimens are known or suspected to have been treated with chemical pesticides, e.g., naphthalene, take necessary measures to protect against harmful exposure.

---

1 & 2 Workspaces for mounting and drying specimens.

POSITIONING THE MATERIAL FOR MOUNTING
Martin Xanthos

Once the initial checks are complete, the prepared specimen, labels, capsules, etc. will need to be positioned on a mount ready for attachment. Some of these steps might require repeating from the preparation for mounting stage.

The plant material
- Specimens should ideally be orientated in the natural growing position, e.g., with the roots towards the bottom, and flowers towards the top

- To allow maximum coverage of all parts, specimens can be positioned diagonally on the mount. In the case of climbing and scrambling plants where orientation is difficult to discern, a circular arrangement is preferable to maximise coverage

- Avoid attaching specimens close to the edges of the mount, keeping at least a 1 cm border around its perimeter. Herbarium specimens are most commonly handled around the edges where poorly mounted specimens could become damaged over time

- For specimens consisting of many tiny individuals, position them with the heaviest individuals at the bottom of the sheet. This prevents the finished mount from being top-heavy

- Tall specimens can be positioned folded over or concertinaed to fit the sheet. Where this is not possible, specimens can be cut at a suitable point and mounted onto a second sheet or more as necessary

Labels and determination slips
- The placement of the label should be carefully considered. By positioning the label in the bottom right-hand corner, this is the easiest and quickest position to consult the text. The label can be positioned to overlap the specimen if space is limited. Avoid positioning over bulky parts of the specimen such as a thick branch to prevent damage to the label

- If available, any determination slips should be attached above the label in date order (earliest first), or if none are present, a space should be left between the label and the capsule (if mounted on the same side) to allow for determination slips to be added at a later point

- Never overlap the label with the capsule

Capsules and jeweller's tags
- Some herbaria have adopted a policy to include a capsule for every specimen, to include loose leaf material for DNA extraction in addition to containing detached parts of the specimen, either before or after the mounting process

- If possible, position the capsule at the middle right-hand side. Heavier capsules can also be positioned at the bottom; this is to avoid the specimen being top-heavy

- The jeweller's tag should never be removed from the specimen and should be attached directly to the mount

TIP

Additional capsules may also be mounted that contain, e.g., small maps, extra labels, which for convenience cannot be mounted separately.

Barcodes and QR codes are used for linking specimens to online data (Diazgranados & Funk 2013). For mounting, these codes are usually positioned along the top or bottom of the mount.

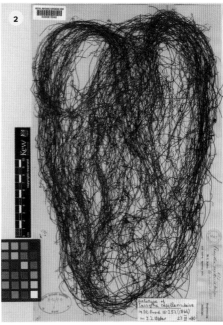

1 Specimens can be positioned diagonally on a mount. **2** Climbing habit of *Cassytha capillaris* arrangement on an herbarium sheet. **3** Stem concertinaed to fit the sheet. Note the capsule position in the middle right-hand side of the mount. **4** Jeweller's tags should remain attached to the specimen.

METHODS AND TECHNIQUES FOR MOUNTING SPECIMENS: GLUING

Martin Xanthos

There are two main methods used for mounting specimens: gluing and strapping. Variations on these methods exist, as well as alternatives to gluing and strapping. The gluing method provides a more securely mounted specimen compared with strapping, but results in the reverse side of the specimen becoming inaccessible.

Gluing the specimen

Prior to gluing, ensure an appropriate sample of plant material has been detached and placed inside a paper capsule. Then apply adhesive by brush or bottle to one side of the specimen, transferring the specimen with forceps or by hand and carefully pressing it onto the mount. Apply minimal amounts of glue and avoid getting glue on the top side of the specimen, which can obscure important plant characters when dried. Small weights can be used to hold the specimen down while labels, capsules, etc. are glued.

If a specimen consists of only one or two flowers, glue down the stem but do not glue down the flower. Instead take a piece of glassine or translucent paper large enough to cover the flowers, apply glue down one side and attach to the mount as protection for the loose flower. This technique is informally known as 'windowing'.

and gluing the rest

The label is completely stuck down unless it overlaps the specimen, in which case, glue the label down one edge only (outer edge is best). Use as little glue as possible to prevent 'warping' and creases appearing on the label as it dries. Thin beads around the edges are preferable. If extra information is found on the reverse side of the label, e.g., maps, only one edge of the label should be stuck down.

Drying

After gluing is completed, the specimen is moved to a designated drying area. Use cardboard backing or other sturdy board to place the first specimen on. Place one sheet of waxed paper then a sheet of blotting paper and wadding on the mounted specimen, and repeat this step until all specimens have been mounted. To maintain pressure on the specimens as the adhesive dries a weight is then placed on top of the bundle, e.g., sandbag and left to dry overnight. Wooden blocks can be used as an alternative drying method (Tucker & Calabrese 2005).

> **TIP**
> The Glass Plate (Singh & Subramaniam 2008) or Dip Method (Tucker & Calabrese 2005) refer to a quick-gluing technique beneficial for rigid and stiff specimens. A flat surface is evenly coated with a film of diluted glue using a brush. The specimen is handled with forceps and carefully placed onto the tray and gently patted down to ensure sufficient glue coverage. The specimen is then transferred onto the mount. The specimen orientation should be visualised in advance of gluing.

1 Applying adhesive to the specimen. **2** Glassine paper (window) mounted over a single flower. **3** Weights hold the specimen down while the adhesive dries. **4** Label overlaps the specimen. **5** Thin layer of adhesive on reverse side of label. **6** A stack of specimens+waxed paper+blotting paper+weight.

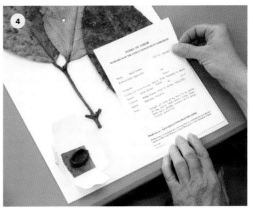

METHODS AND TECHNIQUES FOR MOUNTING SPECIMENS: STRAPPING, STITCHING, WIRING

Martin Xanthos

The main advantage of mounting specimens by strapping is that it is easily reversible, and the reverse side of the specimen is accessible. Stitching is an additional method used alongside other techniques and is especially useful to secure larger pieces of plant material such as fruit or bark.

Strapping

Small strips of tape, 1–5 mm in width, are placed across the specimen at intervals with each end of the strip securely fixed to the mount. The thickness of the specimen determines the width of the tape used, e.g., thin straps for pedicels, thick straps for stems. For bulky stems, the straps can be reinforced with stitching (see Fig. 3).

The tape can be of various types, e.g., cloth, gummed linen and archival self-adhesive tape. Use enough strips to secure the specimen to the sheet but not so many that important plant details are obscured such as flowers, petals or leaf tips. A recommended spacing between adjacent strips is 8–15 cm; 2 lines down: c. 10 mm (Rankin 1992).

Straps should extend c. 10 mm beyond each side of the plant part. Use forceps to tuck in the straps under and around the stems. This minimises movement once the specimen is mounted.

Stitching or wiring

Sewing with a needle and thread is a valuable aid for providing reinforcement of thick stems in addition to gluing or strapping. The thread can be linen, cotton, or thin archival thread as long as it is strong but not too thick. The thread is sewn through the sheet and knotted off on the reverse side with the knot covered with gummed paper to prevent it catching on other sheets when in storage. Stitching is useful for mounting specimens of bark, which may not lie flat on the sheet. Bark is best mounted showing the external side. Consider using a heavier weight card mount for extra support where needed.

Wiring is a more secure version of stitching using non-rusting, corrosion-proof wire e.g., nickel-plated copper wire, two small holes are made on either side of the part to be secured and the wire is threaded through. On the back of the sheet the wire is twisted, folded down and archival tape placed across the wire.

> **TIP**
> A variant of strapping commonly used by herbaria in Japan is the Hotgun Method developed by Dr. H. Kanai while he was the head of Botany at the National Museum of Nature and Science, Tokyo. The Hotgun tip is applied to polyethylene-laminated self-adhesive paper tape and adhered to the mounting paper. The tape width is ideally 5 mm or 10 mm.

1a & b Mounting a specimen using the strapping method and using weights to apply pressure during the drying of the glue. **2** Use forceps to tuck in the straps under and around the stems. **3** Stitching a thick branch to the sheet. **4a & b** Stitch covered with gummed paper on the reverse of the mount. **5** Makino Botanical Garden Herbarium technique using the Hotgun Method.

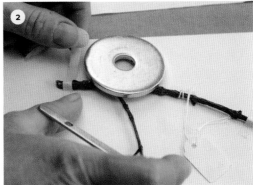

MOUNTING SPECIALIST PLANT FAMILIES
Martin Xanthos

Certain plant groups and indeed whole plant families need special attention in the mounting process due to their unique morphologies and habits. Some of those exceptions are explained (below) in the context of the overall gluing method. If in doubt, consult a specialist regarding the features that are most important to display and which mounting method will best support study in the long term. Mounting practice should also take into account how the specimens are stored within cupboards and the level of handling they will be subjected to.

Reverse mounting technique for flimsy plants, e.g., Balsaminaceae, Hymenophyllaceae

The delicate nature of these plant groups means that once glue is applied to the specimen, it is almost impossible to retain its shape when transferring to the mount. A method known as 'reverse mounting' should be adopted and can be used in combination with the 'windowing' technique if appropriate.

- Arrange the specimen on the mount, detaching delicate parts and placing in a glassine package and capsule, if sufficient material is available
- Place a sheet of waxed paper over the top of the specimen
- Invert the mount+specimen+sheet over so that the waxed paper is now at the bottom
- Remove the sheet and apply glue to the specimen
- Place the mount paper back on the specimen and invert the stack again so that the waxed paper is now on top
- Rub down the specimen through the waxed paper so that the glue contacts the sheet
- Carefully remove the waxed paper and glue any free specimen parts as necessary

Grasses & Sedges (Poaceae & Cyperaceae)

The 'flowers' or spikelets in these families are important for identification. For this reason, a few spikelets should remain free though still attached to the specimen to allow accessibility for study. Achieve this by applying glue to the inflorescence axis and lateral branches. Use a glue bottle, as opposed to a brush.

Ferns

Often important characters for identification are found on the lower side of the fronds, i.e., the sori, venation or pilosity. For this reason, at least one frond should be mounted with the underside facing upwards. Fronds can be considerably large, and one frond can take up several sheets. Successive segments of the frond (i.e., from base to apex) should be mounted in order, with one end comprising sheet 1 following through to the other end comprising sheet 2, 3, 4 etc.

Nepenthes spp.

The underside of the peristome lid is important when identifying species. For this reason, it is best to 'window' the pitcher instead of gluing it down.

Bryophytes & Algae

Bryophyte specimens are not directly mounted onto sheets but are instead placed in packets and subsequently mounted onto a sheet for storage.

Charophytes and seaweeds are prepared for the herbarium at the time they are collected. The sheet that the specimen is pressed on can subsequently be mounted onto the herbarium sheet in the usual way. Specimens not fully adhered to the pressing paper or are too brittle and fragile (as is often the case with charophytes) are placed inside large paper capsules, and in turn fixed to the mount for storage (see The collection of special plant groups, page 38).

> **TIP**
>
> Inflorescences of bulrushes (*Typha* spp.) may 'explode' on the herbarium sheet after a given time (Victor *et al.* 2004). When mounting these specimens place the inflorescence in a polyester wrapping and attach the bottom and top to the sheet to prevent the flowers escaping.

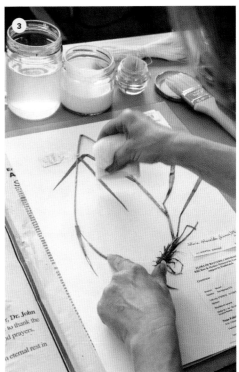

1 & 2 Reverse mounting technique. **3** Use a glue bottle and sponge to secure Poaceae material.

HANDLING
HERBARIUM SPECIMENS
Sally Dawson

Herbarium specimens are valuable sources of information for research, but they are breakable objects, and vulnerable to damage. The purpose here is to understand the risks involved with handling and working with specimens and show how, with care and simple techniques, the risks can be greatly diminished.

The risks to unmounted and mounted specimens

If specimens are handled incorrectly then damage can occur, including tears to mounting paper, specimen parts and labels becoming broken, detached and missing. Leaving specimens exposed can result in pest attack, chemical or substance spills and dust and dirt accumulation. These can occur when specimens are:

- Tipped on their side or turned over
- Held only at the base or with one hand, carried with no cover and no rigid support beneath
- 'Pulled' out from a pile without first removing those above
- Studied in small, limited areas where people, specimens and equipment compete for table space
- Small and delicate fragments, picked up and held in the hand
- Left out of storage for extended periods, especially if left uncovered
- Stored near to food and drink, risking possible liquid or greasy spills and finger marks
- Not repaired at the earliest opportunity after damage occurs and no record of severe damage kept
- Not aligned when added to folders, boxes or shelving
- Added to already full folders or storage spaces

Reducing the risks

The risk of damage is reduced when specimens are:

- Always kept horizontal and facing up (Victor *et al.* 2004; NatSCA 2013)
- Always supported, held on both sides over work surfaces when lifted, covered and placed on boards when moved short distances, put in boxes on trolleys for longer distances
- Removed one at a time from the top of a pile to see those below (Victor *et al.* 2004)
- Studied in spacious areas where they can be spread out or stacked in low neat piles on tables alongside accompanying curation and research equipment
- Small and delicate fragments, picked up using tweezers or pieces of paper slid under for support
- Covered when left out of storage for short periods and put away when work has finished, closing cupboards and boxes (Victor *et al.* 2004)
- Kept away from food and drink
- Mended as soon as attachments fail or parts loosen, small plant parts are put into packets and any severe damage is documented
- Carefully aligned and only added to folders, boxes or shelves with sufficient capacity (Victor *et al.* 2004)

N.B. Wearing conservation quality gloves is advised, as this can protect the specimens as well the user from any chemicals; but check institutional policy as some collections are not suited to glove usage (e.g., National Archives 2022).

Special Collections

- Keep historic, fragile and special collections separate from others or in separate covers. Digitise and control access to view for those specimens most at risk

- Keep hazardous specimens, those with irritant hairs or sharp spines in separate covers. Add warning signs highlighting the danger. Wear gloves to protect your hands. Keep a full list of hazardous taxa

TIPS

- Use a small, duplicate set or synoptic collection of common taxa for demonstrations and teaching. This will help minimise wear and tear on more important specimens
- Do not wear loose clothing or jewellery etc. that could catch on specimens (Timbrook 2014; WAM 2017)

Many herbaria have treated specimens with different chemicals for pests, including mercuric chloride, and it is always advisable to wash your hands after handling specimens.

Good handling.
1 Specimen held on both sides, taken in and out of a folder.
2 Carrying specimens on a supporting board. **3** Spacious work area, board supporting the specimen.

HANDLING ASSOCIATED COLLECTIONS
Sally Dawson

Many collections are linked to herbarium specimens and either kept with them or stored separately and cross-referenced. Letters, photographs, illustrations, or plant parts such as carpological collections, spirit collections or plant products and artefacts in an ethnobotanical collection need similar care when handling to herbarium specimens. They may need specialist treatment.

Spirit collections
- Use safer spirit for study, e.g., 'Copenhagen mix' (with no formalin or toxic fixative), and work in a well-ventilated area
- Wear gloves when handling bottles and working with specimens. Further Personal Protection Equipment (PPE) may be necessary
- Hold bottles upright, by the sides or base not by the top
- Use plastic carriers to transport bottles safely
- Keep spirit bottles standing in drip trays on tables separate from herbarium specimens and away from light and heat
- Use tweezers, needles, pipettes etc. and dishes suitable for working with spirit
- Clear up spills at once
- Replace bottles in a secure store directly after use
- Wash hands after working with spirit

Photographic prints and negatives, illustrations, written documents etc.
- Be aware of the media used before handling. Consult a conservator if media is friable
- Keep items covered with archival polyester to protect the surface when viewed
- Support a mounted object when lifting with two hands, one hand beneath as the mounting may not be secure
- Framed items can be stacked face to face; small, framed work can be carried two at a time, face to face
- If images or written documents are attached to specimens, store them in specially labelled covers for protection

Ethnobotanical collections and bulky/carpological collections
- Storage areas, work surfaces and containers for transporting may need shaped foam and mounts to support and stabilise three-dimensional objects (WAM 2017)
- Work in a team and use specialist trolleys to handle and transport exceptionally large, awkwardly shaped and heavy objects. Plan your route and destination before moving the item (WAM 2017)
- Use both hands to support delicate objects from underneath or from the strongest parts. Never pick up by a weak point such as a pedicel, rim or handle (Timbrook 2014; WAM 2017). Slide a supporting board or tissue paper underneath to help lift very delicate items
- Store and display in clear topped boxes to reduce handling

FURTHER READING
National Museums Scotland (2022);
Natural History Museum (2014).

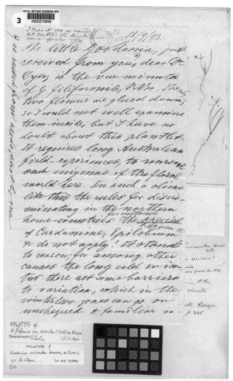

1 Herbarium specimen with illustrations attached. 2 Illustrations drawn directly on the mounting paper. 3 Letter with mounted specimen on right hand edge. 4 Working with spirit material. 5 Holding a basket over a work surface with both hands. 6 Supporting large fruit with both hands.

HERBARIUM ARRANGEMENT
Timothy Utteridge

Herbarium specimens can be arranged following different systems, but regardless of the system used, herbarium curators and researchers, as well as visitors, must be able to locate specimens quickly and efficiently.

Unique and varied systems

There is no widely accepted herbarium arrangement, with different systems used depending on size, history and scope. Herbaria must consider the optimal way to arrange the herbarium collections at family, genus and species level, according to their primary users, workflows and outputs – systematic studies, regional checklists, identification of vouchers etc. Importantly, the system must be unambiguously communicated to aid location of material (Rabeler *et al.* 2019), e.g., see the large posters used on the compactors at KEP and letters on TAN.

At the genus and species levels, systems are often unique to each herbarium, with different combinations of taxonomic, phylogenetic, alphabetic, geographic and floristic arrangements. Special systems can be used developed in different families and genera, i.e., within one herbarium the system below the family level may differ across the collection rather than one single general system.

Systems of arrangement for herbarium collections, especially at family level, are either taxonomic/phylogenetic or A–Z/ alphabetic.

Plant families

Botanists use plant families as a useful way to communicate and understand plant diversity. Families are usually the first taxonomic ranks learnt when undertaking plant identification in the field and herbarium.

Plant families are the most logical higher-level group in which to arrange specimens. Within a phylogenetic system, families are grouped within orders, giving the user an additional layer of information in which to locate the specimens and learn evolutionary relationships.

Phylogenetic systems

Taxonomic and phylogenetic systems use a linear representation of evolutionary relationships. Several systems have been followed, including Engler, Thorne, Bentham & Hooker. Many herbaria using a phylogenetic system are now rearranging their collections to the Angiosperm Phylogeny Group (APG) system (http://www.mobot.org/MOBOT/ research/APweb/).

One advantage of the phylogenetic system is that related families are close together. If there are changes to family delimitation and circumscription, and perhaps its name, then the specimens will remain close to each other. A phylogenetic system would allow such changes to only require relabelling of specimen folders, rather than reshuffling the entire collection to accommodate moving specimens from the old families to the new location. For example, there are several groups, which were split along temperate and tropical distribution, such as the Theophrastaceae and Myrsinaceae, which are now placed within an enlarged Primulaceae, or woody members of the Verbenaceae now placed within an enlarged

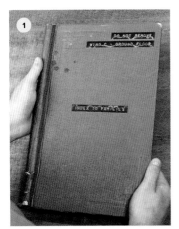

Family arrangements need to be clearly communicated to users. **1** At RBG Kew a red folder is placed on each floor of the collection. **2** The Kew folder lists the name, number and location of each family. **3** In the KEP herbarium, the families are arranged phylogenetically with large posters giving the family name, number and location. **4** TAN herbarium is arranged A–Z with large letters on the cupboards indicating the sequence position.

Lamiaceae; as such, in the latter case, a phylogenetic system means that the families after Lamiaceae (families L–U in an alphabetic system) did not need to be shuffled along to accommodate the ex-Verbenaceae specimens.

The physical layout of a phylogenetic system allows users of the herbarium to place themselves physically within the plant kingdom, and are thus powerful tools when teaching students evolutionary relationships.

Users will need to understand plant relationships and the published system on which the arrangement is based. It can be confusing, and time consuming, having to learn a new system. While the APG system is now being adopted more generally across herbaria, there are still several different evolutionary classifications in use.

A–Z/alphabetic systems

Arranging families alphabetically is a very straightforward arrangement for all users and visitors. An example of a major herbarium using an A–Z arrangement of families, genera and species is the national herbarium of Indonesia (Herbarium Bogoriense, BO). It is very common for herbaria to place the genera in alphabetical sequence and is probably the most convenient and straightforward to learn. However, with the proliferation of new molecular approaches, many generic limits are being changed and collections can quickly become outdated if curation is not maintained. In medium to large-sized families, this may result in major reshuffling if there are significant changes. See, for example, new generic limits in Acanthaceae, Gesneriaceae or Convolvulaceae.

Type collections

Type specimens are the nomenclatural reference point for every plant name and should be highlighted in the collection, e.g., many herbaria keep types in special red folders. Researchers and visitors will often want to examine and request type specimens above other material. Type specimens are important and should be cared for and protected appropriately. At Kew, types are incorporated within the collection, but other herbaria keep type (and historic) specimens in separate rooms for additional security (e.g., BO and CANB); however, for research, it is more efficient to study species if types are incorporated within the general collection.

Herbaria will use different geographic boundaries to arrange their specimens. **1** The Kew geographic arrangement – the same colours are used on regional labels on generic folders. **2** Being an herbarium with a largely Malaysian/SE Asia collection, the KEP geographic arrangement is focused on the Asia region – note that outside Asia, specimens are 'Exotic'.

Genera and species:

Many herbaria use taxonomic and/ or phylogenetic lists of genera, e.g., Dalla Torre & Harms (1900–1907), with updated generic relationships being revealed with the publication of DNA phylogenies (Baker *et al.* 2022). It can be

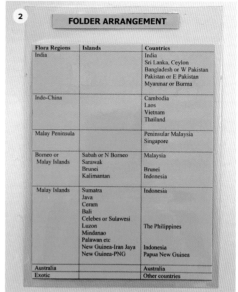

2 FOLDER ARRANGEMENT

Flora Regions	Islands	Countries
India		India
		Sri Lanka, Ceylon
		Bangladesh or W Pakistan
		Pakistan or E Pakistan
		Myanmar or Burma
Indo-China		Cambodia
		Laos
		Vietnam
		Thailand
Malay Peninsula		Peninsular Malaysia
		Singapore
Borneo or	Sabah or N Borneo	Malaysia
Malay Islands	Sarawak	
	Brunei	Brunei
	Kalimantan	Indonesia
Malay Islands	Sumatra	Indonesia
	Java	
	Ceram	
	Bali	
	Celebes or Sulawesi	
	Luzon	The Philippines
	Mindanao	
	Palawan etc	
	New Guinea-Iran Jaya	Indonesia
	New Guinea-PNG	Papua New Guinea
Australia		Australia
Exotic		Other countries

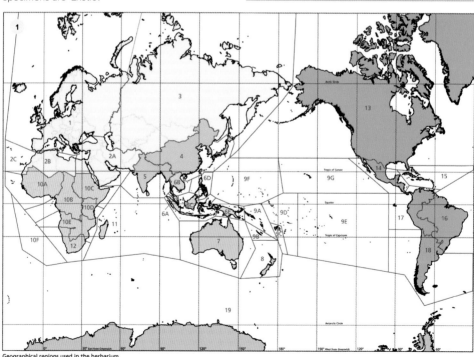

Geographical regions used in the herbarium

most straightforward to arrange genera and species in an A–Z/alphabetic system. Generic-level phylogenetic systems have been proposed, such as the linear Annonaceae sequence (Chatrou *et al.* 2018), now adopted at Kew after a comprehensive re-curation of the family. Phylogenetic systems at genus and species level are very useful in herbaria undertaking naming projects, and it can be useful to check adjacent groups if identifications are problematic as evolutionary-related taxa are often morphologically similar.

Specimens not formally identified: Herbaria may have specimens still waiting formal naming to species. Those named only to genus are placed at the end of the appropriate genus in 'Species indet.' folders; sheets named only to family can be placed at the end of the generic sequence in 'Genus indet.' folders. Very rarely, there may be sheets yet to be assigned to family (often difficult to name as lacking key information) – these sheets are usually at the very end of the entire herbarium sequence in a 'family indet.' section. These specimens should not be discarded! The 'indet.' folders are often a treasure trove of exciting new records or undescribed species.

Physical v. virtual systems
With the advent of mass digitisation, specimen images will be widely available, allowing researchers to view collections online. It will remain important to undertake physical and digital curation as part of the current workflow in herbaria is to browse the collections enabling associate specimen selection. Automated warehouse storage and/or digital views (due to the smaller image size of specimens on a screen) hinder this workflow (Hardy *et al.* 2020). Physical specimen access will remain fundamental to plant diversity research.

TIPS
Rearranging collections
The Kew Herbarium was completely rearranged to APG, but not all herbaria will have the resources to undertake such a task and prioritising families that are a particular strength of an organisation will be strategically better. However, it is important to estimate what resources are needed, such as additional staff, stationery, equipment etc., and the financial consequences, before starting a reorganisation.

- Reorganisations always take longer than estimated and require a lot of people power

- During a move, some 'holding' space is needed in the herbarium to accommodate collections after they are removed from the sequence but before being reinserted in the new place

- Although freezing and spraying the cupboards with insecticide makes the whole process longer, it is very effective at containing pest outbreaks in the short and medium term

- If freezing specimens during the process, factor in time for the specimens to warm up before putting them away, otherwise condensation may form

The Angiosperm Phylogeny Group, usually abbreviated 'APG', is a consortium of researchers publishing updated classifications of flowering plants based on the most up-to-date phylogenetic information. The current classification is APG IV published in 2016, and a Linear representation of APG III that can be used in arranging herbaria was published by Haston *et al.* (2009). Family limits, orders, characters and notes on the different genera are regularly updated on the Angiosperm Phylogeny Website (Stevens 2001 onwards).

INCORPORATION
MATERIALS
Nina Davies

It is useful to have ready to hand equipment needed for incorporating specimens into the collection. This will depend upon whether you are incorporating flat herbarium specimens or bulky collections.

Materials for organising and incorporating mounted specimens

Equipment for herbarium specimens
A medium soft pencil (HB, 2B or F), a rubber and sharpener are important for the curation of herbarium collections. Pencil annotations can be easy edited and erased from the specimens and to add useful information to specimens when organising collections to a sequence such as the generic number or geographical sequence.

Archival quality pens are useful for annotating genus information on folders; for example, at the Kew Herbarium a thick black pen (0.8 mm nib) is used for folders and a thin pen (0.3 mm nib) is used for determination slips that are added to specimens during the identification process. However, printed labels can also be used for annotating folders, e.g., family and geographical label information at Kew. These should be attached securely with glue.

Different types of covers are needed for curating herbarium specimens, for example genus folders, which are usually card, species folders, which are usually paper, and specialised folders for type or historic material. It's useful to have new genus and species folders nearby while incorporating specimens in order to split folders as they fill up and to avoid any damage from overfilling.

Materials for organising and incorporating bulky specimens

Equipment for bulky specimens:
Specialised boxes are useful for bulky material such as carpological collections. Other bulky collections might need to be placed in particular folders, for example, types, bulky palm and cacti collections.

1 Example of specialised boxes used for storing and incorporating bulky collections such as palms. **2** Examples of the materials for organising and incoporating mounted specimens. **3** Examples of specialised boxes (glass-topped) for storing and incorporating carpological specimens.

HERBARIUM SPECIMENS
Nina Davies

Specimens must be filed into the collections in the correct place, otherwise they may be lost for a considerable period of time, especially in larger herbaria. This section outlines the main procedures to correctly incorporate specimens in herbaria, including special collections.

Organising mounted specimens before incorporation

Mounted specimens are organised in the following order: family, genus, geographical area (if applicable), species, sub-species and/or varieties (and rarely forms).

Generally, the following guide details how to organise specimens before incorporation:

- Gather the appropriate curation materials before incorporation

- Arrange the newly mounted specimens in a logical sequence corresponding to the herbarium arrangement. This will save time as groups of specimens can be incorporated at once rather than going back and forth from collections. Ensure specimens are correctly handled throughout the process

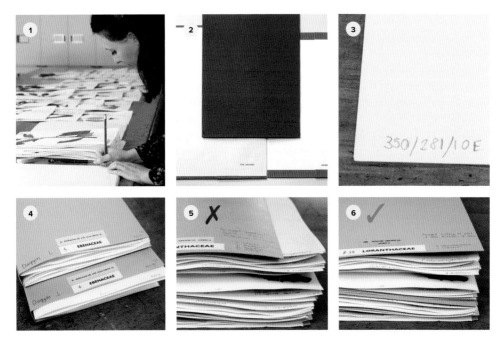

1 An appropriate workspace is needed to spread out and organise specimens. **2** Examples of type folders from different herbaria. **3** For large herbaria it can be useful to annotate family no., geography etc. on the specimen in pencil before incorporation. **4** At Kew: the genus is written in black pen (archival) on bottom left of the folder and the species is pencilled on the bottom right corner of the folder. The family name and geographical information is in the centre of the folder. Species folders are annotated in a similar way in pencil on the bottom right. Multiple folders covering a species in a single geographical area are given running numbers in pencil, e.g., 1/2, 2/2. **5–6** Species and genus folders should not be overfilled to avoid damage from specimens overhanging the edges of the folders.

- If there is a list available with the geographical arrangement, this can be pencilled onto specimens before incorporation. An online search, gazetteer or map can be useful for finding additional geographical information not given on the label, for example, where the collections are split into smaller regions below the country level

- Organise the specimens according to family/genus/species depending on the specimens you are working with, e.g., mixed families or a box of Rubiaceae – for phylogenetic arrangements use the index to find the family number for the specimens and pencil this onto the sheet, or in an A–Z order for alphabetic arrangements

- At this point, if there are specimens labelled as types, these should be placed into a type folder (often coloured)

Procedure for incorporating mounted specimens

Herbaria differ in how specimens are incorporated into the collections, and it is important to find out what the correct method for this is at the herbarium; for example, several specimens of the same species are usually placed together in a species folder, but sometimes specimens might be placed in their own folder, such as historical collections or types.

Now that the mounted specimens have been organised, it's time to incorporate them into the collections.

1. Locate the correct folder corresponding to family, genus, geography and species, e.g., *Coffea arabica* from Ethiopia (Rubiaceae).

2. Open the folder and carefully lift each species folder containing specimens until you arrive at the correct folder (*arabica*).

3. Incorporate this specimen.

Species named to sensu lato
If species are named in the broad sense, i.e., 'sens. lat.' ('s.l.'), these should be incorporated in a separate species folder to the species, sub-species and/or varieties.

Specimens doubtfully named
Any specimens that are doubtfully named (I.e., have '?', 'cf.', 'aff.', 'sp. near' before the specific epithet) are best placed in a separate folder after those with the same epithet that are definitively named; these specimens are often of interest to taxonomists because of morphological differences and worth identifying as such.

Species only named to genus
Specimens named only to genus should be kept in a separate folder marked 'sp.' (singular) 'spp.' (plural for species) at the end of the sequence. If a specimen is determined as 'sp. A', 'sp. B' etc. or 'sp. 1', 'sp. 2' etc., this indicates that either the specimen has been assigned to a taxon awaiting formal naming or some initial sorting of the specimens prior to revisionary work.

> ## TIPS
> - Where there is a large number of specimens to incorporate, organise many at once into the correct sequence. Even if these end up temporarily back in bundles or boxes, they can be incorporated in focused sections, e.g., at family or generic level
>
> - For type specimens, annotating the outside of a closed folder with specimen information enables users to quickly check material, especially where there are multiple types

ASSOCIATED COLLECTIONS
Nina Davies

Incorporating exotic taxa

The recommendations for filing specimens considered as exotic, such as cultivated or weed taxa, may vary in different herbaria. Introduced weeds and naturalised species are often included with the main sequence. Garden escapes can cause confusion as they may not have been noted as such by the collector and are often not included in floristic or monographic accounts. Cultivated taxa are generally easier to note and deal with. If the collections are geographically sequenced, the cultivated folders can be placed at the end and clearly marked as such.

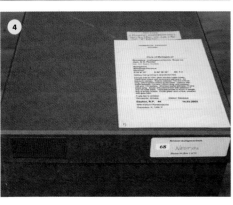

Incorporating returned loans

When returned loans are newly received at the herbarium, check the material to see if new determinations have been added to the specimens while on loan. If any of the new determinations are of newly described species yet to be published, often written as 'in. ed.', consider which name to incorporate this under. It might be best to file specimens under their previous name until formal publication, after which the collections can be rearranged according to this.

Incorporating bulky material

Bulky specimens need special attention for incorporation, but the same general principles apply to these collections as with herbarium specimens. The varying sizes of the boxes or bulky folders that specimens are stored in can make it tricky to organise collections if there is limited space. Again, it is important to find out the arrangement of the collections and if there are any indices to help guide curation. Ideally, separate bulky collections are stored as near to herbarium specimens as possible, for example, the carpological collection, Palms.

All herbarium bulky material in boxes should have a label attached to the outside detailing key specimen label information. Type specimen information should be boldly annotated here, for example, in red.

Considerations for incorporating spirit and ethnobotanical collections

Spirit collections should be housed in a separate room due to the hazards associated with this collection. However, the same considerations over the collections arrangement and storage space must be considered before incorporating collections. These collections can also be arranged numerically, which is better for using the space to the maximum potential, grouping jars of a same size together. Here, information on the bottle number record changes with taxonomic revisions, but the physical collection remains static.

The arrangement of ethnobotanical collections is complex and requires separate considerations for the curation of objects compared with herbarium material (Salick *et al.* 2014). Objects are extremely variable in type and size compared with herbarium collections, therefore it is important to consult literature, and experienced curators and researchers before incorporating and storing these collections.

> **TIPS**
> - When creating labels for boxed herbarium material, add the relevant curation information at this stage to help with incorporation, for example, genus number, geographical code etc.
> - When incorporating spirit collections, keep in mind accessibility – do not layer the bottles too deep

1 Spirit collections – numerical sequence.
2 Carpological collection showing the use of space. **3–4** Carpological and palm box labels detailing specimen information including type annotation. **5** Palm boxes among herbarium specimen boxes in the collections.

CONDITION CHECKING

Donna Young

The current state of a specimen or collection can be assessed with a condition check, which can inform recommendations for treatment, storage, and use. These checks can be used to prioritise and calculate resources needed to stabilise, care for, and maintain collection areas or individual specimens. Keeping condition check records (ideally with photographs) allows collections to be monitored and provides an audit of changes over time. Condition checking is important to ensure long-term preservation, future access and use of a collection.

1 Pest-damaged specimen and frass. **2** Torn margins on mount larger than the genus folder. **3** Detached specimen and loose straps. **4** Dirt migrated onto mount along right-hand edge (open folder side) – possibly via unsealed cabinets. **5** Mount margins and annotated text at risk of further damage through handling.

When to do a condition check

Acquisition – ideally prior to, or when specimens are acquired.

*Specimen loans** – assess the risks associated with damage during postage, unpacking and handling. For example, is the material securely mounted, and the mount stable enough to support the specimen?

Destructive sampling – before and after specimen material is removed.

*Display** – assess the risks involved in the display and transport of specimens. How is the material to be displayed? E.g., flat in a case, upright on a stand, framed? Is the specimen loose or particularly fragile? Is it at particular risk of damage from high UV?

Routine checking – part of general collection management task. E.g., during documentation or imaging, where individual sheets can be assessed, and appropriate decisions made regarding their condition.

Routine monitoring – random sampling from collection areas, assessed on a rotation basis.

Areas of concern – more frequent checks may be carried out on fragile and/or historic material, and vulnerable collections prone to infestation, e.g., *Taraxacum* or Fabaceae.

Post major incident recovery – following a potentially damaging event such as flood, fire, or discovery of pest infestation.

*A condition check should also be carried out afterwards, before specimens are incorporated back into the main collection.

Things to record in a specimen condition check

- Signs of insect damage or mould
- Loose specimen material at risk of detachment. E.g., original adhesive deteriorated, straps loose or insufficient
- Loose or detached labels
- Label deterioration. E.g., iron gall ink acidity damage, fading from light exposure, tears or brittleness
- Torn, damaged, or unsupportive mounting paper
- Non-archival (often deteriorating) materials present, e.g., Sellotape, cellophane, Post-It Notes, biro text
- Pins used to attach labels and/or specimens (may be rusty, loose, sharp)
- Surface dirt
- Stains and residues from pesticides

Collection condition reports may also include an assessment of the herbarium collection storage, including the materials used to directly house the specimens, e.g., genus folders or boxes, as well as shelving and cabinets.

> **TIP**
> Condition surveys are a tool of best practice in collection management. Doing a sample-based condition check on a representative group of specimens may be the most effective means of gathering information in a large collection.

FURTHER READING
Collection Trust (2022; Spectrum 5.1: Condition checking and technical assessment).

REPAIRING AND CLEANING

Donna Young

Specimens may require repairs and cleaning to stabilise and preserve them, enabling future use. The following instructions provide a best practice approach for dealing with these collections. Any repairs should be documented.

Handle with caution

Historic herbarium sheets may have been treated in the past with insecticides, e.g., mercuric chloride, and must be handled carefully. Stains, sometimes grey or metallic, may indicate insecticide use, but many are not easily visible without UV. It is safest to assume all historic material will have been treated. It is advisable to wear a barrier cream, such as Derma Shield, or wear (powder-free) nitrile gloves when working on historic herbarium sheets. A professional museum or archival conservator can be consulted to advise on repairing and cleaning of specimens.

Cleaning

Dirt and dust may accumulate on the mounting paper, and the specimen itself, due to poor storage. Atmospheric dirt such as soot can migrate into folders, especially those holding bulky specimens or buckled mounts. Use a soft brush to remove surface dust and dirt, taking care around delicate areas of the specimen. Do not use water as it can consolidate dirt. The mount paper can be cleaned with a plastic eraser, using gentle strokes away from the specimen, taking care around pencil annotations. Cut the eraser at a sharp angle to use closer to the specimen. An eraser putty can be used to clean large paper areas. Gently remove eraser dust with a soft brush, sweeping towards the edge of the mount. Remove insect frass and debris with a soft brush. Annotate the sheet, dating when the specimen was cleaned of frass so that any new infestation is evident. Try to avoid inadvertently removing small seeds or other detached fragments – retain these in a capsule and attach to the mount.

Securing specimen and labels

Secure larger detached or loose specimen parts by adhering paper straps or stronger pre-gummed linen tape. Cut strips to a suitable width – to secure stems, cut strips roughly as thick as the stems. Secure by tucking and pressing strap edges close to the specimen using a small spatula or closed forceps.

Labels directly attached to specimens can become damaged over time during examination, and labels attached with pins risk tearing or rust damage. It is advisable to remove these labels and reattach using Japanese tissue corner tabs or hinges.

If there is insufficient room for all the labels, place them in an annotated capsule and attach this to the mount. When imaging, place all labels alongside the specimen.

Repairs to the mount

Poor handling and cramped storage can cause tears along the edges of the mount.

Larger than standard sheets can be trimmed if they have sufficient space between the specimen, labels, and the edge of the sheet. Ensure to leave enough margin for safe handling.

Repair small tears by pasting acid-free tissue over the back of the tear, or by using purpose-made archival document repair tape or mending tissue.

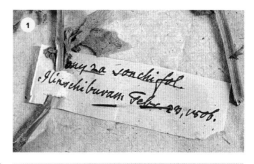

1 Original label attached directly to specimen and label attached to mount with pin. **2** Mercuric chloride staining on herbarium sheet. **3a** Japanese tissue tabs adhered onto reverse corners of ink script label. Tab edges can be then adhered directly to mount. **3b** Label ready for reattachment to new mount sheet. **4** Removing label attached directly to specimen. **5** Erasing surface dirt. **6** Brushing eraser dust and dirt away from specimen.

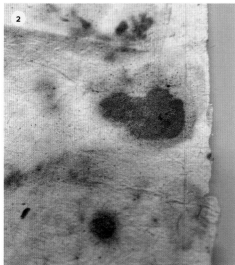

FURTHER READING

The following organisations can provide direction for further collection care advice and guidance:

NatSCA (Natural Sciences Collections Association)

ICON (The Institute of Conservation)

SPNHC (Society for the Preservation of Natural History Collections)

REMOUNTING SPECIMENS
Donna Young

There are circumstances where herbarium sheet repairs and cleaning may not be sufficient to facilitate its long-term use and preservation. The decision to remove a specimen from its original support and remount the specimen should be made on a case-by-case basis.

Remounting

The original mount and means of attachment may be of historical significance (watermarks, distinctive strapping, etc.) and curatorial assessment essential before any modification. In these circumstances the specimen and its mount can be attached to an archival back-mount using corner tabs of Japanese tissue. Any annotations on the reverse of the original sheet should be imaged for reference. Removal and remounting an entire specimen, and its associated data, may be necessary to stabilise it for future use. All materials used should be archival quality, with all actions reversible and fully documented. Curatorial assessment may be required to assign original labels to correct new mounts. A professional museum or archival conservator can be consulted to advise on remounting and the preservation of specimens.

1a Attaching strap securely by tucking close to specimen. **1b** (detail). **2** Removing softened adhesive tape after sheet gently humidified. **3** Reattaching original label with Japanese tissue corner tabs. **4** Label annotated on both sides remounted using Japanese tissue hinge. **5** (reverse detail). **6** Remounted herbarium sheet.

Reasons to remount

Inadequate support – due to mount being lightweight, torn, weakened or pest-damaged.

Obscured features – due to specimen originally being poorly pressed, mounted, or overcrowded on mount. Remounting will allow rearrangement and if necessary, multiple consecutive sheets (cross-referencing sheets and photocopied labels attached to each one).

Mixed species or collections – mounted either deliberately or accidentally. These sheets can make correct taxonomic and geographical storage difficult. However, mixed sheets may be preserved together for historical importance. In this instance, printed images of these sheets can be inserted elsewhere in the collection for cross-referencing.

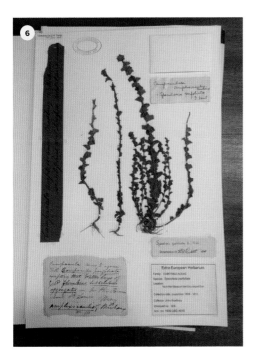

Removal of specimen and labels

- Stitches and straps securing a specimen to a sheet can be cut with a scalpel

- Paper or linen straps may be lifted from a specimen using forceps – use a soft brush to gently moisten a small area with water to ease away pieces that remain adhered

- Some glued specimens may be gently released from a sheet using a spatula eased carefully under small areas

- Water-soluble adhesives can be softened using a hand-held humidifier or after placing the sheet on a damp absorbent pad for a few hours. In both cases, labels should be removed first to protect unstable inks. Specimens needing relaxing prior to rearrangement and re-pressing can also follow this procedure

- Insoluble adhesives and animal glues may require advice from a professional conservator

- Firmly attached specimens can be cut away, leaving a 1 cm margin around to indicate the original mount, and attached to a new sheet

- All labels should be transposed to the new sheet, along with the specimen

- Any annotations directly made onto the original mount should be carefully cut out and attached as labels to the new sheet

> **TIP**
> When more than one herbarium sheet is being worked on, extreme care should be taken to ensure original labels are kept with, and reassigned to, their correct specimen.

DATABASING AND DIGITISATION
INTRODUCTION
Sarah Phillips, Laura Green, Marie-Hélène Weech and Clare Drinkell

Digitisation is the process of converting analogue information and making it available in a digital format. This section will cover the process of selecting specimens, assigning unique identifiers (barcoding), transcribing specimen labels into a database and creating one or more high-resolution photographs or scans of the labels and plant specimen, resulting in a virtual herbarium specimen which can be shared online. It will also cover some databasing and imaging software. Digitisation enables a broader audience access to herbaria through online platforms in a form that can be easily analysed and shared, facilitating curation and research.

Digitisation: an explanation
Revealing an herbarium through digitisation can act as an inventory to better understand the complete holdings of a collection, increasing accessibility and potential. For example, it can be a useful resource for the discovery of new species, with online information supporting national environmental policies (Canteiro *et al.* 2019). The user community of a virtual herbarium expands with digital collections, providing a platform for novel partnerships, collaborations and a wider range of applications/analyses, e.g., studying phenological responses to climate change. While limited, computer algorithms enable taxonomic identification through artificial intelligence techniques such as machine learning, by inspecting thousands of digitised specimens.

Databasing: unlocking the narrative
Capturing label data into a database structures the information to be searchable and analysed. Implementing common data fields, standards and formats allows users to interpret the data more easily and aggregators to bring datasets from other herbaria together providing an even richer dataset.

Key information includes field collection information (who, when and where), taxonomic name and collector number, and references to ancillary collections.

Imaging: a picture is worth a thousand words
An herbarium specimen digital image is a high-resolution photograph or scan incorporating the plant material, labels, barcode, colour chart and scale bar. Access to images allows remote online users to know what collections are held by an institution, what they look like, and utilise specimens for virtual research. Access to original label information allows users to interpret the written information for themselves, see the original handwriting and put the data into historical context. These images also aid in processes such as identification and morphometric analysis through viewing morphological characteristics. However, users should appreciate the limitations to these processes where details, such as hairs or interior floral parts, cannot clearly be observed. Even if such limitations of these images require the physical study of a specimen, digital image catalogues can help users select specimens (Enghoff 2019).

1 Digital image of K000921396 *Deguelia decorticans* showing the specimen, label, barcode, measure and colour chart. This species was first identified as new to science by examining digital and physical specimens.
2 The digital image of an herbarium label allows users to interpret the written information for themselves.

VALUE OF SPECIMENS

Each herbarium specimen, once digitised, remains fundamentally important as a voucher to validate scientific observations, and should not become superfluous. In addition to fundamental use cases such as DNA-analysis, chemical analysis, dissection and other physical analysis, the 'sensory turn' of a specimen is an important part of explorative investigation. The object is important culturally, historically and as the source of the digitised specimen, remains the very foundation of the herbarium and associated research and curation.

LIMITATIONS

The 'digital divide' excludes certain user communities lacking adequate access to technological infrastructure, making the physical collection essential for particular user groups. Bulky specimens such as fruits in carpological collections cannot easily be digitally imaged in 2D and require investment in terms of time, money and skills to capture digital imagery suitable for scientific analysis.

> **TIP**
> Try to embed digitisation into current curation practices where resources allow. For example, database all new accessions and specimens going on loan. Develop workflows to capture data digitally as soon as possible from the point of collection in the field.

WORKFLOW FOR DIGITISATION: PRE-DIGITISATION CURATION

Sarah Phillips, Laura Green, Marie-Hélène Weech and Clare Drinkell

Specimens need to be prepared for digitisation, and the number of tasks needed will depend on the state of curation of the collection and whether the specimen is going to be databased, imaged or both. It is important to factor in time for this stage in any digitisation project proposal.

Specimen preparation

The first stage in any digitisation workflow is specimen preparation, e.g., specimens may need to be mounted on herbarium sheets, collection labels printed, sheets may need to be repaired or the storage information updated.

Specimen selection

Digitisation rates can be significantly impacted by the time it takes to select the material out of the collection. It is advisable to design projects in which you digitise larger units of the collection in one effort rather than selecting individual specimens that are likely to be scattered around the collection. The most efficient workflow will usually be to select material in the order in which it is stored in the herbarium, e.g., working through sequential blocks of cupboards or compactors.

When selecting specimens from the collection, it is helpful to leave markers/tags in the collection to indicate material has been removed. Such placeholders should include contact details of the person who has the material, date of removal and details of the specimens such as number of folders and taxa. These placeholders will aid the return of the material to the correct place in the collection.

If specimens are being moved or transported to a digitisation station, then they should be placed in boxes of an appropriate size and moved on a trolley to avoid damage when in transit.

Barcoding specimens

Each specimen should be given a unique identifier, which acts as a key that connects the data and image created to the physical specimen. It is standard practice to use barcode labels, which can be acquired through an external supplier or printed in-house using a barcode font and archival grade paper and ink. The barcode can then be entered into the database record using a barcode reader. Each barcode must be unique, so it is common practice to use the herbarium acronym or code at the start of the barcode followed by a number series with a standard number of digits. Most commonly used are one-dimensional linear barcodes (iDigBio, 2015) but two-dimensional barcodes are also in use. The number should also be printed in a human-readable format and the institution's name should be stated.

An ode to the barcode

Most institutions use self-adhesive barcode labels, but it is worth noting that if using plastic barcodes, the adhesive is variable and untested for conservation soundness. For best practice it may be necessary to insert a protective layer to avoid acrylic adhesive migration or print the barcode on archival paper as part of the collection details label. Some particularly historic specimens may not lend themselves to having barcodes directly attached to the herbarium mount. In this case it is possible to attach a cover sheet overlay of paper-stock known as 'Heritage wood-free' for

1 Example of a placeholder in the herbarium indicating a specimen has been temporarily removed from the collection. **2** Example of a barcode. **3** Barcode reader. **4** The preferred position of a barcode is the top or bottom edge of the herbarium sheet. **5** An historically important specimen with barcodes attached to an overlay of Heritage wood-free paper.

applying the barcode. Attach the paper to the reverse top of the specimen sheet using wheat starch paste and then fold over the full length of the herbarium sheet (see Fig. 5).

WORKFLOW FOR DIGITISATION: ELECTRONIC DATA CAPTURE

Sarah Phillips, Laura Green, Marie-Hélène Weech and Clare Drinkell

Data are transcribed into a database or Collections Management System (CMS). The system that is chosen will be dependent on the needs of the institution, available resources and IT infrastructure. Choices include in-house development or out-of-the-box solutions such as Microsoft Excel and Access, and commercial/consortium systems with different business models, such as Brahms, EarthCape, Symbiota, Arctos and Specify. It is recommended to gather requirements for the institution's digitisation project and then assess systems against these.

Minimal data capture

It may be necessary to capture the information that is written on the containers and folders in which the specimens are stored if this information is not also on the specimen labels themselves. Specimens are usually stored in folders, which may list the country, genus and the name of the species the specimen is stored under within the collection. These minimal data are needed to enable curators to find and retrieve the specimen from the collection and should always be recorded within the digital specimen data. It is recommended to capture this information into the CMS at the same time as barcoding the specimens and therefore may be considered a pre-curation step.

Label transcription

Label transcription involves capturing information regarding the collection that is found on the specimen and can occur pre- or post-imaging. This can be done either directly into the CMS or using an intermediate database and importing into the CMS at a later stage. If transcription is completed after imaging, this can be done from the image, allowing collaborating researchers in other countries or online volunteers to transcribe the specimens remotely. It can often be useful to include some minimal data fields to assist transcribers such as country and taxonomic name.

Ideally all the information on the herbarium specimen labels should be transcribed, but if the digitisation project involves large volumes of collections to digitise, it is often appropriate to prioritise the remaining information that is captured, such as the following fields:

- Unique ID or barcode
- The most recent taxonomic identification
- Collector
- Collector number
- Collection date
- Country
- Locality information
- Altitude
- Plant description
- Habitat description
- Vernacular names
- Coordinates, if present on the sheet
- Type status, if relevant

Level of data captured

Even within the same institution, the level of label transcription can differ from one specimen to another. Different digitisation projects may demand and fund a different level of data capture depending on the end use of the data. Biodiversity Information Standards (BIS), historically known as Taxonomic Databases Working Group (TDWG), have set up a Task Group to define a specification for the Minimum Information about a Digital Specimen

(MIDS) with the aim of harmonising the information to be expected from different levels of digitisation (Hardisty & Haston 2021). MIDS is divided into several levels, with MIDS Level 1 the basic level of information about a specimen. We recommend that all digitisation projects should aim to at least capture information to MIDS Level 1.

> **TIP**
> Online volunteers can be asked to help capture specimen label data. There are many crowd sourcing platform solutions available, such as ALA DigiVol, Les Herbanautes, Notes from Nature, DoeDat and Zooniverse.

A specimen with the components highlighted

Locality description: Detailed information on where the specimen was found allows researchers to map the distribution of the species and also to return to the same site in the future.

Coordinates: Captured with GPS receivers as standard practice, allow spatial data to be incorporated into GIS computer applications.

Habitat description: The physical and biological conditions under which the individual was found provide information on the necessary conditions required for the species to grow and reproduce.

Elevation: Also often known as altitude above sea-level. Provides another dimension of spatial data for GIS and habitat analysis.

Plant description: Information about the plant as it was when it was collected. Features of the plant that cannot be collected or that get lost upon pressing help to identify the specimen later on.

Collection date: Gives biologists studying reproductive behaviour vital clues to the times of the year a plant can be expected to be in flower and fruit. Also useful in evaluating likelihood of current presence given land use in assessing extinction risk.

Determination slip: Updated scientific name, which may differ from the initial determination on the specimen label.

Taxonomy

Determiner

Herb. Hort. Kew.: this clearly indicates the herbarium in which the specimen is housed

Barcode

Capsule

HERB. HORT. KEW.

ROYAL BOTANIC GARDENS KEW

K001502470

PERU

ANNONACEAE   Dpto. Madre de Dios

Oxandra   mediocris   Diels
det. P. Maas '92

Prov. Manu; Parque Nacional Manu
Rio Manu; Pakitsa Station; first bend downriver from camp.

11°50' S, 71°16' W.
High forest on low floodplain.

Alt. 350m.
Tree, 15m; short horizontal branches; fruit yellow-green, not ripe.

Robin B. Foster   12636   19 December 1988
& Severo Baldeon
BIOLAT   SMITHSONIAN INSTITUTION (US)

Oxandra acuminata Diels
det. Leo Junikka 2007
University of Helsinki, Finland

WORKFLOW FOR DIGITISATION: SPECIMEN IMAGE CAPTURE

Sarah Phillips, Laura Green, Marie-Hélène Weech and Clare Drinkell

Herbarium specimens can easily be imaged to a high quality using a camera and copy-stand with flash or LED lightbox lighting. Image files should be saved as non-proprietary and lossless file types. Several key components should be included in all images, including a scale bar, colour chart, data labels and a unique identifier in the form of a barcode.

Image quality

Digitisation requirements should be ascertained using measurable and standardised guidelines such as Federal Agencies Digital Guidelines Initiative (FADGI) (Rieger 2016) or Metamorfoze (van Dormolen 2012). Both of these guidelines provide tiered performance levels, from the highest preservation grade quality to more basic digitisation quality, drawn from different tolerance levels met over a variety of image quality criteria. It may not always be possible to produce digital images to the absolute highest preservation grade level; however, the highest quality levels achievable, given the budget available, should be aimed for. Generally, the recognised standard for creating herbarium specimens to a suitable resolution is a sampling frequency of 600 PPI (Häuser *et al.* 2005) while ideally meeting the criteria for the appropriate performance level in the chosen guidelines.

Imaging equipment

Many suitable imaging solutions have become available in recent years (Sweeney *et al.* 2018, Davis *et al.* 2021). In addition to choosing a camera that allows high-resolution image capture, it is important to choose appropriate lenses, copy stands/mounts, lighting and tethering software to achieve a high-quality image. A camera with an image sensor of 80 megapixels or more paired with a high-quality lens will enable production of images of 600 PPI or more. However, lower-cost DSLR cameras and lenses can still be used to create images of a lower PPI that still allow for label information to be captured, even if more detailed characters cannot. Computer specifications, additional equipment such as barcode readers and the space needed for a suitable workflow should also be considered. Tethered imaging is strongly recommended. This

can be achieved through software such as Adobe Lightroom, CaptureOne, or the software provided with your camera if this allows for tethered imaging.

Components to include in the image

The following components should be included in an image(s) of an herbarium specimen:

- Barcode – a unique identifier; also used to link the specimen image to the digital record
- Colour chart (also known as colour targets or colour checkers) – used to verify the colour accuracy of the specimen, which can be adjusted post-processing

- Scale bar or ruler – allows downstream users to measure plant parts accurately from the image. It is common to include the institutional logo and short copyright statement. Scale bars and rulers are therefore commonly designed by the institution and then printed and laminated
- Labels – all collection labels, determination labels and annotations should be clearly visible. It may be necessary to create more than one image to capture all this information if labels overlap
- Specimen – all plant material should be captured where possible. If a label is obscuring key characters, additional images should be taken with the label folded back

Imaging unusual specimens

Not all herbarium specimens will come mounted on standard herbarium sheets. Material from families such as Pandanaceae and Arecaceae is often so large that it cannot be mounted in this way. In addition, carpological material is often very bulky and identifying key characters to capture can prove difficult. Different approaches to digitising these types of material need to be explored to gain the best images of the specimen.

TIP

Investing time at the beginning of your digitisation project to research the most suitable equipment and workflows for your needs will greatly reduce the amount of time that is required for quality assurance and repetition of efforts later on.

1 Example of an imaging workstation. 2 Camera set-up for imaging a specimen. 3 Imaging a bulky palm specimen, using a depth-of-field measure.

WORKFLOW FOR DIGITISATION: IMAGE PROCESSING AND STORAGE
Sarah Phillips, Laura Green, Marie-Hélène Weech and Clare Drinkell

Specimen images should be captured and stored at the highest archival quality available and can therefore be very large. Careful thought and preparation are needed in order to store and maintain them securely.

File naming and formats
All specimen images should be captured in raw file format and then converted to Tag Image File Format (TIFF). These non-proprietary lossless image files are archive quality master images that are suitable for long-term preservation. It is widespread practice to name the file under the specimen identifier, usually a barcode, thus enabling linking of the image to the specimen and its digital record, e.g., K000921396 (Nelson *et al.* 2012). TIFF files can be named using a barcode scanner. It is also possible to implement software within the digitisation workflow that can read barcodes within the image to automatically rename the file to the barcode value (Nelson *et al.* 2012; Sweeny *et al.* 2018).

In addition to the master TIFF file, it is common to create derivative files for online publishing or for sharing and distribution; common formats include JPEG and JPEG2000.

Image storage

It is vital that the safe storage and future digital preservation of master images is considered. Images should be backed up securely with multiple copies in separate locations to mitigate hardware failure and disaster. File integrity tests should be performed on archiving (e.g., check sum tests) and successful retrieval of files from archival storage should be tested. The UK based Digital Preservation Coalition (DPC) provides a useful handbook for guidance (DPC 2015).

> **TIP**
> A suitable naming convention where additional images of a specimen are required is to add a suffix to the file name, e.g., K000134587.TIFF, K000134587_a.TIFF, K000134587_b.TIFF.

Digital Asset Management system

Images should be stored alongside image metadata (a set of data that describe and give information about the file). This includes who digitised the specimen, when it was imaged, file size, what equipment was used and the image capture settings. Some of this information will be automatically embedded into the file on creation. RBG Kew stores and manages within a separate Digital Asset Management System, a central repository for the institution's images. The Collections Trust have produced a guide to support cultural heritage organisations in understanding how to approach the integration of Digital Asset Management (DAM) into their existing collections practice (Poole & Dawson 2013).

1 Image processing.
2 View of a specimen image on screen.

WORKFLOW FOR DIGITISATION: EXCHANGING AND MAKING DATA ACCESSIBLE

Sarah Phillips, Laura Green, Marie-Hélène Weech and Clare Drinkell

To maximise the research impact of digitised collections, the data and images should be made widely available. Where possible, data should be released under FAIR principles promoting findability, accessibility, interoperability and reuse of digital assets (Wilkinson *et al.* 2016).

Making data available online

Many institutions have the capacity and infrastructure to create their own portal to give external access to their collections, such as the Royal Botanic Garden Edinburgh Herbarium Catalogue (https://data.rbge.org.uk/search/herbarium/), collections management system (CMS) suppliers will often provide their own solutions that could be assessed for suitability. However smaller institutions can still provide data through biodiversity data aggregators such as the Global Biodiversity Information Facility (gbif.org), iDigBio, Europeana and Jstor Global Plants. Data and images are shared with aggregators through data exchange standards such as Darwin Core Archive (DwC-A) and the Access to Biological Collection Data (ABCD) Schema.

Licensing use of data and images

Institutions retain copyright for the data and images they create, but it is recommended data are made available under an open licence to promote free and open access. Many funding bodies and national governments are now making this a

Royal Botanic Gardens Kew | Plants of the World Online

HOME   ABOUT   HELP

Family: *Fabaceae* Lindl.

Genus: *Delonix* Raf.

Delonix regia (Bojer ex Hook.) Raf.

Delonix regia is a distinctive tree with large, bright red flowers. The genus name is derived from the Greek words *delos* (meaning conspicuous), and *onyx*, meaning claw, referring to the appearance of the spectacular flowers. The tree is commonly cultivated in the tropics and subtropics, including Madagascar, for its ornamental value, but is under increasing threat in its natural habitat due to habitat destruction.

requirement when obtaining funding for digitisation and research (Wilkinson *et al.* 2016). RBG Kew currently distributes specimen records and images under Creative Commons licence CC BY 4.0, which allows for sharing and reuse if clear attribution is given (Creative Commons 2020).

Data that cannot be shared

Some specimens might have restrictions on their use including online access. This can be due to many reasons such as agreements made with other institutions or countries or due to Red List assessment category. Ensure a copy of any collection agreement is maintained and any restrictions on reuse recorded. Institutions will develop their own policies regarding release of data. For example, specimen records of very rare or critically endangered species might be excluded from external release, or detailed locality data may be removed before publication (Morton 2020).

> **TIP**
> Get advice and keep up to date with data exchange standards and vocabularies by participating in the Biodiversity Information Standards community (www.tdwg.org).

The stable and precise referencing of specimens used in scientific studies requires the use of specimen identifiers that are globally unique, consistent and reliable; for example, Kew has adopted the Consortium of European Taxonomic Facilities (CETAF) stable identifier system (Güntsch *et al.* 2017). This gives Kew the ability to provide a Persistent Identifier, Persistent URL and Stable URI for each specimen, e.g., http://specimens.kew.org/herbarium/K000757492. These stable identifiers are increasingly used for referencing specimens in publications and data portals allowing improved tracking of use of specimen data.

Plants of the World Online featured taxon page of *Delonix regia*, *Plants of the World Online* | Kew Science.

RECOMMENDATIONS
Sarah Phillips, Laura Green, Marie-Hélène Weech and Clare Drinkell

In an increasingly digital world, there is a growing expectation from users that all collections should be digitally available. While it is extremely challenging to find resources and funds to meet this expectation, institutions should work towards embedding digital curation alongside physical curation.

Make it part of your every day
Creating, maintaining and updating digital specimen records should be as important as the curation of the physical specimens; ideally, digital workflows should become embedded into curation practices. Herbaria should be working towards "Digital First" workflows with digital records created at or soon after the point of collection and digital records exchanged between institutions along with physical specimens.

As time goes by...
Institutions should invest in digitisation training for staff to enable them to manage imaging facilities, design efficient workflows, and develop and implement institutional standards. Digitisation is often project-funded and without staff continuity, without clear documentation and ongoing training expertise can be lost when project funding ends. Dedicated staff can also keep up to date with international data and imaging standards.

A stitch in time saves nine
Quality Assurance (QA) procedures should be included when designing workflows. Providing feedback to transcribers and spotting issues with images early reduces the amount of data cleaning or reimaging needed later. QA also highlights any inconsistencies and areas where procedures or protocols need to be updated. Data should be validated where appropriate at the time of data capture and procedures implemented to check each specimen record has a digital image and vice versa, ideally before the specimen is returned to the collection.

Specimens in synergy
Enriching specimen data through activities, such as georeferencing and linking herbarium specimens to other resources including other collections, field images, species descriptions, Floras, molecular data, protologues, collector notebooks and archives, will open them up to a multitude of other uses and users.

> **TIP**
> Many institutions and Collection Management Systems are already including Optical Character Recognition (OCR) in their workflows (Drinkwater *et al.* 2014). This can work well for typed written labels but still needs manual intervention to check and make sure the information goes into the right fields. As technology improves, further automated processes will be included in digitisation workflows.

1 Embedding digital curation into the daily workflow. **2** Electronic data capture, Harapan Forest, Sumatra (Indonesia). **3** Capturing electronic field data. **4** Data born-digital access during fieldwork. **5** Point map on a smart phone.

DESTRUCTIVE SAMPLING
Toral Shah

The use of molecular, cytological, phytochemical, palynological and anatomical data for research in plant systematics and taxonomy is growing rapidly, leading to an increase of sample requests for the necessary material. Destructive sampling should not be done without permission, and guidance from a curator should be followed.

General advice
Principles with general recommendations for sampling are given in Rabeler *et al.* (2019).

- Institutions should produce guidelines and policies for the removal of material from specimens to uphold their scientific integrity
- Before sampling, consult repositories to understand current availability of material, for example, the Kew DNA Bank lists sampled DNA available
- Sampling requests should be first sourced from home countries and institutions; if material is not available, requests can be made to larger herbaria
- Samples should not be taken from type specimens or specimens of historical importance unless special permission is granted
- Samples should be taken from specimens in good condition and only if adequate material is available – if possible, loose material should be taken from capsules
- Avoid removing the smallest or largest leaf from a specimen which may be important for identification
- Try to avoid material that has been collected using the Schweinfurth method – a technique whereby plants are soaked in 70% alcohol; such specimens are often dark brown or black and this may interfere with analytical methods

- Always attach an appropriate label to the specimen indicating the following information: type of material removed, who it was removed by, when and the purpose of removal

Sampling for molecular studies
Samples between 1−5 g or a c. 2 x 2 cm size of representative leaf from herbarium specimens should be taken if fresh leaf material is unavailable. As above, use material from packets, avoid Schweinfurth material or leaves with glue attached as compounds may interfere with DNA extraction protocols. If leaf material is not available, other plant material may be sampled (e.g., stem, root, bracts). Check with a curator before removal. Do not remove reproductive characters such as flowers or fruits unless plentiful.

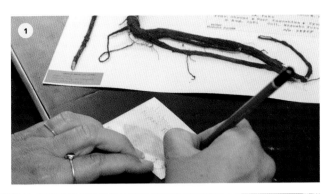

1 Placing sample in packet and labelling with pencil. **2** Sampling using forceps, slips and barcodes, as well as a laptop to record specimen/voucher information directly into a database or spreadsheet. **3** Slips to be attached to specimens to show sampling information. **4** Sampled specimen with completed and stuck-down sampling slip.

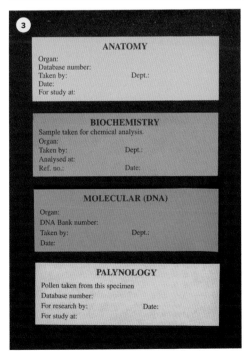

ANATOMY
Organ:
Database number:
Taken by: Dept.:
Date:
For study at:

BIOCHEMISTRY
Sample taken for chemical analysis.
Organ:
Taken by: Dept.:
Analysed at:
Ref. no.:   Date:

MOLECULAR (DNA)
Organ:
DNA Bank number:
Taken by: Dept.:
Date:

PALYNOLOGY
Pollen taken from this specimen
Database number:
For research by: Date:
For study at:

Sampling for pollen, spores and palynology

Pollen or spores from herbarium specimens are useful for plant taxonomy and systematics (see Jarzen & Nichols 1996; Jarzen & Jarzen 2006).

- In angiosperms, pollen is usually produced in the anthers; in gymnosperms, pollen is in cones or similar structures
- Check the specimen bears enough flowers and/or buds to allow removal of a sample, ensuring the scientific value of the specimen is not diminished
- Ensure the specimen has fertile flowers with developed staminate parts (check if flowers are unisexual!). Mature flowers or buds are for best pollen samples but ensure anthers have not dehisced; specimens collected just before anthesis are usually ideal
- Use magnification to see inside flowers and check for pollen grains

Carefully remove the pollen-bearing structure and place in small envelope. If bracts are obstructing the flowers or buds, carefully remove them and return to the specimen capsule.

Collection of sporangia is more difficult as spore-bearing organs are not always recognisable. For ferns, consult a fern manual or specialist to identify position of the indusia and the location of sori (Tyron & Tyron 1982; Tyron & Lugardon 1991); you will need magnification to observe the sori and indusia position. It is important to identify if the material was collected with mature sporangia.

Sampling for phytochemistry (Cook *et al.* 2021)

Ensure selected material is air-dried and not collected under the Schweinfurth method. The latter method may

> **TIP**
> Use Glassine or Cellophane envelopes for sampled material as these prevent material from sticking. On the outside of the envelope, use archival permanent marker or pencil to write the plant name and specimen details.

have caused leaching of chemical compounds of interest. The quality of the phytochemical analysis depends on multiple factors:

- The drying time and drying method of the herbarium specimen
- The level of compounds at collection of the specimen
- If the specimen incurred further deterioration during storage

The quantity of dried material may depend on the phytochemical method being implemented. Consult the chemist or relevant protocol before sampling. Similarly, the part of the plant requiring sampling may depend on the study in question. Again, consult the chemist before sampling.

Indicate in the notes attached to the sample or collect a sample of the glue or adhesive used to secure the specimen to the herbarium sheet, as this may be needed as a control in the phytochemical analysis.

Sampling for anatomical studies

Anatomical studies may require different plant parts including leaves, stems, flowers, and individual floral organs. It is important to consider all the required parts of the specimen before sampling to avoid removing large amounts of material from any one specimen. Check if there is loose material in the specimen capsule and consult the spirit collection prior to sampling.

Avoid sampling for specimens that have been collected using the Schweinfurth method as the alcohol may have modified structures such as waxy layers or hairs.

Leaves

Remove a representative leaf size and shape from the specimen. Do not remove the smallest, largest, or youngest leaf. If the leaves are relatively small, take the whole leaf. If the leaves are large, use a scalpel to carefully cut a section from the leaf. Cut about 1.5 cm wide, from the margin to the mid rib if possible. For monocotyledons, try to include the leaf base and apex. For studies looking at leaf surface, ensure there is no glue or adhesive on your sample.

Stems

For wood anatomy try to select the oldest twigs of at least 1.5 cm thick. Using a sharp blade cut a piece (1–)2–3 cm length.

Flowers

Before removing flowers or buds from the specimen, ensure there are sufficient flowers available to prevent diminishing the scientific value of the specimen for future research.

Sampling from spirit collection

- Spirit collection material can be used for numerous different disciplines including scanning electron microscope (SEM) analyses, anatomy and palynology
- As with herbarium specimens, spirit collections should only be sampled if adequate material is available
- Spirit collection material is most often sectioned and dissected for anatomical research

---

TIPS

Sampling seeds: In some instances, seeds may be required for phytochemical or anatomical study. First, ensure the herbarium specimen bears enough seeds or fruits to warrant removal from the specimen. Usually mature and well-formed seeds are needed.

For phytochemistry, usually the size of the seed determines how many you will need. If seeds are greater than 1 cm, then 2–3 seeds will be sufficient. For smaller seeds, more may be required.

For anatomy, study may be of the entire seed morphology or just the seed coat; therefore, 1–3 seeds would suffice.

LOANS
REQUESTING A LOAN
Sara Edwards

Although the digitisation of specimens is useful for research and may often be adequate, loans of herbarium specimens between institutions continue to be essential for many research projects, especially where microscopic morphological characteristics need to be examined. Loans are beneficial for herbaria to get specimens identified and annotated by specialists and cited in papers. Particular care should be paid towards the safe keeping and longevity of specimens during the loaning period.

Dear Keeper of the Herbarium,

Loan of specimens to [name of institution where the specimens will be sent for study]

I am writing on behalf of [name of researcher], a [research position e.g., PhD student] based at [e.g., the Royal Botanic Gardens, Kew] to request a loan of specimens of [a brief summary of taxa e.g., genus and species, and geographical range] from [loaning herbarium e.g., Herbarium Bogoriense].

[Name of researcher] is carrying out a taxonomic revision of [xx] as a part of his PhD and he would like to examine vegetative and reproductive structures of [xx] in detail. Herbarium Bogoriense is the only herbarium that can provide specimens of these species. Please kindly provide the specimens to complete his PhD.

Here is the list of requested specimens

[give as much detail as possible, provide guidance to the sending institution what material should be prioritise for loaning. Institutions rarely lend all their holdings and knowing what the researcher has already available can help the lender identify the most useful material to send e.g., the researcher already has on loan /or had studied material from MO and NY Herbaria]

[As necessary include the following detail]
Geographical range
Genus
Species names
Synonyms {for complex groups}
Collector name
Collection number
Date of collection
Barcode of specimen

All specimens will be handled following standard procedures at [loaning institute] and carefully examined whilst they are in the herbarium. Please do not hesitate to provide any suggestions on the specimen. The materials and institutions will be acknowledged in the publications as the source of data.

Yours sincerely
[signature]

[Name — The loan request should come from someone with the authority necessary to accept a loan on behalf of the institution]
Head of Collections
[Herbarium name]

Example of a loan request letter which should be emailed as a pdf document on a branded letterhead template.

Loan requests

Loan requests should be sent to the Director or Curator of the herbarium. The request should include who is applying for the loan and why, their affiliated institution, a list of species (with synonyms for complex species groups), the geographical areas and whether permission for destructive sampling is requested. If types or specific specimens are required, their collector numbers, dates and locations should be included. Loans are usually given a return date of 6 to 12 months. Many herbaria post their loan policies online.

A loan request may be rejected if the specimens are already in use, if the number of specimens is considered too large, the species requested are very rare or it would be easier for the researcher to visit the collection in-situ. For the loaning institution, it is good to check whether there are issues that might endanger the specimens, for example, environmental or security concerns (Paine 1992; Victor 2004; Rabeler *et al.* 2019).

Check list for sending and receiving loans, whether outgoing from your institution or incoming from another institution

- Before entering or leaving an herbarium, the specimens should be treated to reduce the risk of spreading insect pests between institutions, e.g., frozen for 3 days at −30°C, 7 days at −25°C or 14 days at −18°C. (see: Pest management)
- Check the numbers of specimens being sent or received tallies with the paperwork or specimen list. Write a condition report of specimens, entering problems noted before sending on loan. If damage to specimens during the loan is noted, contact the loaning institution to inform them about it. If you are sending the loan, keep a record of the damage and make repairs before sending (see: Record Keeping)
- If specimens received are undamaged and match the paperwork, inform the loaning institute that it was received in good order
- Give each loan a unique loan reference number and keep records of paperwork

Incoming loans (from another institution)

- When receiving a loan, write the running sheet number and the unique loan reference number in pencil on each sheet. When returning, remove these marks. This makes the loan easily identifiable. Alternatively, if the sheets have barcodes, these can be added to the loan record
- Attach a label to the bundle of specimens which includes a summary of the loan record with details such as the name and address of the loaning institute, the date it was received, taxa included in the loan, geographical area and who the loan is for. It can be useful to colour-code labels, e.g., red for loans from other institutions, blue for loans from your institution and green for specimens being sent as a gift. A copy of the paperwork should be kept with the loan
- Keep the loan in safe, secure and pest-free conditions
- Always gain permission from the owner before undertaking any dissections or destructive sampling, imaging the specimen, sharing with a third party, transferring the loan onto another institution, gluing on labels or identification slips or carrying out repairs on the specimens
- When returning the loan, where possible use the original flimsies and packaging material (Collections Trust 2022)

PREPARING A LOAN
Sara Edwards

When selecting specimens for a loan, many institutions keep several specimens behind so their work, e.g., identifying incoming accessions, can continue. Many herbaria have a policy of only sending up to half of their specimens at any one time, which protects against loss of entire collections when in transit and on loan. If a researcher would like all the material suggest sending it in two batches, the second to be sent on receipt of the first.

Selecting specimens for an outgoing loan (from your institution)
Things to consider when choosing which specimens to send:

- Full extent of morphological variation and unusual characters are represented
- Full geographical range is represented (if requested)
- Are the specimens historically important (if so, perhaps send newer specimens)?

- Are the specimens firmly mounted (if not, get them repaired)?
- Are there any restrictions prohibiting sending the specimen on loan to a third party?
- Consider sending relevant unidentified specimens, as the specialist researcher may identify these and there may even be unknown species (new to science) among the collection

1 Loan stamp with unique loan reference number and a running sheet number helps keep track of loaned herbarium sheets.
2a & b. 2a A hanging label for attaching to the loaned bundle of specimens. A copy of the label can be placed in the collection as a record of the loan until its return.
2b Hanging labels for incoming borrowed loan, specimen on loan and outgoing loan. **3a & b** Spirit material being prepared for shipping.
3a Muslin thoroughly soaked with Copenhagen mix is used to carefully wrap the specimen, ensuring it is completely covered, to help the specimen remain moist during transit. **3b** The stages of preparing a specimen for transit (from right to left) – remove specimens from storage jars, wrap in Copenhagen mix soaked muslin and place in transit container; print x2 labels, one for the transit container, the other for the bag; double bag to prevent leaks.
4 Bulky or carpological collection ready for shipping.

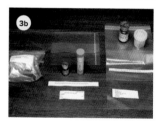

Processing an outgoing loan

- It is useful to digitise specimens being sent on loan, especially types. If this is not achievable, a record and number of specimens named to family / genus / geography / collector number wherever possible should be kept and a copy sent with the loan (see: Record Keeping)

- To ensure that the specimens are clearly labelled as coming from your institution, a stamp can be useful to mark each sheet. The unique loan reference number and a running sheet number can be written in pencil on each sheet. This is helpful for counting returning loans and identifying which loan the specimen is from

- Hanging labels can be put in the herbarium collection stating where the loaned specimens have gone, and details of the loan

- Check if any of the species requested are listed on the Convention on International Trade in Endangered Species (CITES). If yes, complete the relevant CITES label to accompany the loan for CITES-listed institutions or arrange for the correct import and export permits for non-CITES-listed institutions (see: Legislation)

- Ensure that the loan is securely packaged

- A copy of the loan regulations accompanies the specimens in transit, which can be signed and returned by the receiving institution

- When the specimens have returned from being on loan, check the sheets for new determinations and update any database records. Put the specimens back into the collections under the new names if you agree with them

Spirit, bulky/carpological and ethnobotany collections
Melissa Bavington and Sara Edwards

Where possible, if the loan herbarium sheet has ancillary spirit, carpological or bulky, or ethnobotany collections, it is useful to send the complete specimen set.

Spirit collections
All loan material should have a substance information sheet detailing the substances, associate risks and explaining how to handle and work with the specimens. All loan material is kept in Copenhagen mix, which does not contain formaldehyde (see: Processing Spirit Collections). Spirit material should be drained of all fluid and wrapped in muslin. The specimen and muslin are then soaked in Copenhagen mix and drained. The specimens are placed in wide-mouthed plastic bottles for transport and the specimen ID and loan transaction number added to internal and external labels. The specimens are then double bagged ready for shipping (Bentley 2007).

Carpological or bulky collections
These are removed from their permanent boxes and put into travelling cardboard boxes. The specimens should have padding in the boxes to minimise movement. Copies of the labels should be put in the travel boxes and originals left in the herbarium boxes, so the specimen can be easily reunited with its box on return. Where possible, attach jeweller's tags with taxa, collector name and number to the specimen, so it can be easily identified if separated from the label.

Ethnobotany collections
Much the same as the carpological collections; however, also check artefacts for animal-derived parts and check if the items are CITES-listed.

MATERIALS FOR TRANSPORTING HERBARIUM SPECIMENS
Clare Drinkell

The transportation of herbarium specimens for gifts or loans makes them highly vulnerable to damage or loss, so particular care is needed in the packing material and labelling of shipments to ensure suitable protection (Woodruff 2008). This includes mounted and unmounted specimens and any ancillary collections, e.g., ethnobotany material (see also Specimen Exit and Shipping).

Equipment for packing herbarium sheet specimens

The main hazards to specimens in transit are poor handling, exposure to unsuitable environmental conditions and the danger of insect or fungal attack within the package. The following materials and processes will aid safer transit.

1 Incoming parcel.
2 Parcel being prepared for shipment.

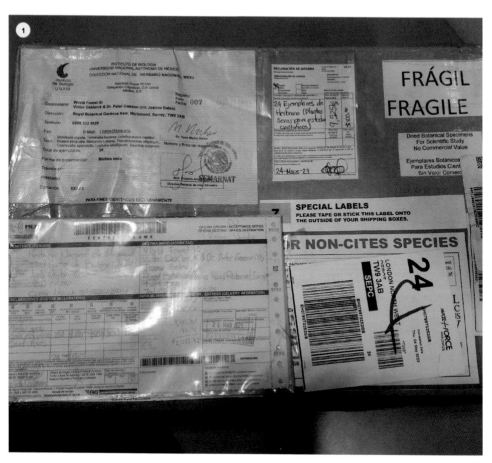

Equipment
- Herbarium-sized cardboard with corrugates running lengthways for maximum strength
- Acid-free tissue 'flimsy' double herbarium sheet size, folded in half, or newspaper, preferably unprinted, to catch loose material
- Strong cardboard box, 1–2 cm larger than herbarium sheets to allow for additional padding
- Cardboard pieces, or any inert padding. Avoid starch padding, wool or other material that can attract pests
- String
- Bitumen paper or other medium to heavy-duty /water-resistant Kraft Union paper
- Parcel tape
- Freezer

- Optional – stapler, loan stamp and ink
- For bulky material – small cardboard boxes, padding
- For spirit material – lint-free muslin, wide-mouth plastic container and lid, jeweller's tag, plastic bag
- For outgoing gifts or duplicates – 'ex' herbaria label
- For incoming specimens – hanging label with reference information

Shipping mounted and unmounted herbarium material
- Select specimens for transit and prepare documentation for dispatch (see shipping and loans sections)
- Line a cardboard box with padding
- Enclose each specimen in a flimsy or newspaper. Take extra care with unmounted material and staple newspaper if necessary
- Bundle specimens tightly enough to limit movement into a 10 cm pile and enclosed within two cardboard sheets
- Make sure the bundle is balanced level on top and secure with string using an herbarium knot
- Wrap the bundle in heavy-duty paper and seal with parcel tape
- Repeat with another bundle if necessary, include paperwork, ensure contents are securely padded, seal and label the box ready for dispatch

TIPS

Upon arrival, incoming specimens should be frozen for 3 days at –30°C, 7 days at –25°C or 14 days at –18°C. While incoming loans from other herbaria are in your institute it is best practice to keep them bundled within cardboard sheets and a hanging label to identify the consignment.

Gifts or duplicates should be sent to other institutes with 'ex' herbaria labels.

HERBARIUM HIGHLIGHT:
RIO DE JANEIRO BOTANICAL GARDEN (RB)

Rafaela Forzza

Jardim Botânico do Rio de Janeiro, Brazil, is an important botanical research institution with The Herbarium Dimitri Sucre (RB) founded by the naturalist João Barbosa Rodrigues in 1890. The first collection was a private donation by Emperor Dom Pedro II of about 25,000 specimens.

Key points

- RB comprises c. 850,000 land plant, fungi and algae specimens
- The collection is completely digitised and is managed using the JABOT system developed by JBRJ staff
- RB holds c. 15,000 types and all are available online
- The RB data are publicly available in several online biodiversity portals (e.g., the RB institutional database JABOT, as well as the Reflora, SiBBr and GBIF portals)

Arrangement

The collections are organised alphabetically across two floors in the herbarium building: the dicots families of angiosperms from A to M are stored on the first floor; the remaining families of dicots, monocots, gymnosperms, ferns, lycophytes, bryophytes, algae, fungi, lichens and additional collections (fruits, spirit, wood and ethnobotanical) on the second floor. The angiosperms are organised according to Cronquist (1988) and are now gradually being reorganised to APG. The gymnosperms are organised according to Christenhusz *et al.* (2011); ferns and lycophytes follow PPG I (2016); bryophytes follow Söderström *et al.* (2016); and algae follow Guiry & Guiry (2017).

Important collectors

Barbosa Rodrigues, Brade, Campos Porto, Duarte, Ducke, Dusén, Fée, Glaziou, Hatschbach, Kuhlmann, Martius, Rizzini, Schwacke, Spruce, Sucre, Ule.

Scope and research

Current activities are focused on the Brazilian flora. Around 20,000 specimens are added to the collection each year, half of which are collected by Jardim Botânico do Rio de Janeiro (JBRJ) staff with partners from around Brazil; the remainder is exchanged with herbaria worldwide.

The Herbarium RB leads the Brazilian Flora and Reflora Virtual Herbarium projects (http://reflora.jbrj.gov.br). It also supports a wide range of research at JBRJ, especially taxonomy and systematics, conservation assessments, and seed banking and other conservation initiatives.

Visits and loans

Specimens are sent on loan and c. 550 scientists visit the Herbarium each year from other institutions to study our collections for their research. Visits are requested through rb@jbrj.gov.br. We provide c. 2500 loans and c. 10,000 of exchange material each year with herbaria around the world.

Herbarium Dimitri Sucre. **1a** Front view of the Herbarium building. **1b** Internal view of the entrance hall with interpretation. Collections: **2** Fungi. **3** Spirit. **4** Ethnobotanical. **5** Fruit and bulky material. **6** Algal.

BUILDING AND ENVIRONMENT

SPECIMEN STORAGE
ENVIRONMENTAL CONDITIONS
Nina Davies

The ideal environmental conditions for herbarium collections are explained below.

Temperature

Insect pest activity and breeding are reduced at temperatures between 5 and 15°C, and increased above 20°C (Pinniger 2015). Considering this, and a compromise to ensure an ideal working environment, it is recommended that herbarium collections are maintained between 16 and 18°C.

Avoid fluctuations in temperature to reduce the likelihood of moisture formation which encourages mould growth. Where specimens have been treated with poisons, e.g., mercuric chloride, specimens are at less risk from pests at higher temperatures.

Relative humidity

Storing collections at 40–60% RH and avoiding dramatic fluctuations within this (±5%) is recommended (Pinniger & Lauder 2018). Lower humidity levels reduce the risk of damage to specimens caused by pests, fungi or mould growth. Data loggers are useful for monitoring temperature and relative humidity.

Depending on the local climate, it may be necessary to install central heating and/or air-conditioning to control the temperature throughout the year, and possibly de-humidifiers to reduce humidity. In tropical regions, if de-humidifiers cannot be obtained, locating the herbarium on upper floors may help in reducing the humidity.

Ventilation

Air-conditioning systems in closed spaces help to maintain a steady temperature and relative humidity for collections.

However, in climates where conditions cannot be controlled over 24 hours, good air circulation may be better, reducing the likelihood of dramatic fluctuations in conditions (Townsend 1999). To prevent the entry of insect pests, windows and external doors should be fitted with draught excluders and, in the tropics, with insect-proof screens.

In conditions where the atmosphere within an herbarium has a high concentration of insecticidal, fungicidal or insect-repellant vapour (e.g., naphthalene), staff should work in separate rooms. A fume cupboard should be installed if the staff are to handle any volatile substances, e.g., in the spirit collection.

Light

It is important to ensure adequate lighting. Specimens should be kept in cupboards/boxes or underneath cardboards when not in use to avoid loss of pigmentation from visible light (Purewal 2019).

Pollution

Herbarium collection spaces should be cleaned on a regular basis to avoid specimens being damaged by pollutants such as dust, mould and soot. Opening windows and doors can increase the risk of pollutants entering a collection's space from the outside environment. However, this is minimised by keeping specimens in suitable storage, limiting opening windows or using a closed collections environment with air conditioning (NPS Museum Handbook 1999).

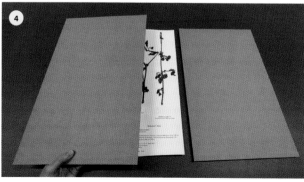

1 Data loggers monitoring the temperature and humidity in the climate-controlled collections space in Wing E, Kew Herbarium. 2 Good air circulation in Wing A, Kew Herbarium. 3 Soot on a *Nepenthes* specimen. 4 Covering specimens with cardboards for protection while temporarily out of the main collection.

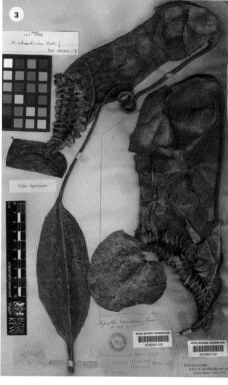

FITTINGS AND FIXTURES
Clare Drinkell

The choice for storage solutions is dependent largely on budget, available space, climatic and geographic conditions. The purpose is to minimise environmental damage to herbarium specimens when in storage, and to enable efficient access to the collections when required for study. Any storage solution should consist of inert and archival materials (Bedford 1999; Moore *et al.* 2019).

Storage options

Boxes
Made from acid-free high-density board (2250 micron) and covered to reinforce with a strong, water-resistant coated buckram cloth. Interior lined with archival paper. Boxes should be at least 2 cm deeper and wider than the herbarium sheet, and approximately 50 cm in height, and fitted with a drop front opening, a built-in card frame and a finger pull.

For special groups such as Palms, or carpological / bulky specimens (see: Processing unmounted specimens).

PROS
- Easy for moving specimens from storage to the working area
- Can be used with fixed or compactor open shelving
- Drop front opening for easy handling of specimens
- Effective for buffering temperature and RH fluctuations
- Can limit damage to specimens from dust, light, flood, fire, and pests

CONS
- Can be more susceptible to environmental damage than cabinet storage
- Specimens are less easy to access than those stored in cabinets

> **FURTHER READING**
> Kurz *et al.* (2021)

Cabinets
Wood or metal cabinet with a tight closing door. Cabinet shelving should be at least 2 cm deeper and wider than the herbarium sheets and shelves approximately 15 cm apart. Doors can be sealed with Plastizote foam seal and provided with a front label slot. Care should be taken over choice of woods, paints or other products such as glues which might emit harmful gases and cause damage to specimens (Timbrook 2014).

PROS
- Well-fitted cabinets will offer maximum protection from environmental damage such as dust, light, flood, fire, and pests
- Wood cabinets are especially effective for buffering temperature and RH fluctuations

CONS
- More expensive than open shelving, requiring more maintenance
- Any cabinet material in proximity with specimens should be free of acids and solvents

1 Storage box open. Note the hinged lid and interior lining. **2** Storage box approx dimensions 20 x 30 x 50 cm (dimensions dependent on herbarium sheet and shelving sizes). Note the front card frame and finger pull. **3** Wood cabinets. **4** Metal cabinets. **5** Showing a range of storage possibilities. Storage boxes on open shelving, with trolley for moving boxes. **6** Compactor storage.

High-density mobile storage units:
Mobile systems, or compactors, consist of shelving, drawers, cabinets, or racks mounted on wheeled carriages running on tracks. Consult guidance from a structural engineer for herbaria in seismic risk areas, and for floor loading capacity.

PROS

- Compactor units can allow for twice or more storage space than conventional fixed shelving
- Flexible storage options

CONS

- Restriction on access on the number of users at any one time
- Increased the risk of physical damage to the specimens through the movement of the compactors

WORKING SPACE
Clare Drinkell

Examining specimens

An active herbarium demands generous working spaces for laying-out material. A suitable area within easy access to the stored herbarium specimens is ideal for creating an efficient and functional workspace. This is either integrated within or immediately adjacent to the storage area, dependent on factors such as available space, light and temperate.

The sorting table should be large enough to spread out specimens for study, be a comfortable height for standing, and have suitable lighting for examining specimens. Extra space for equipment and temporary storage, and additional desk space for sitting at a computer and microscope is essential.

Fitted furnishings are recommended for the workspaces and the floor should be a seamless, hard flooring. A moveable trolley work surface is useful to minimise the transporting of specimens from storage to table workspace.

Reduce risk

The handling of specimens is arguably the point at which the material is most vulnerable to damage and avoiding risk is key to a well-designed work area. Along with training on handling material, users should minimise the movement of herbarium material from storage to mitigate risk (Ertter 1999; Lundholm 2019).

Keep it clean

There should be no food or drink or living plant material in any area with herbarium specimens. The entire space should allow for easy cleaning and pest activity inspection.

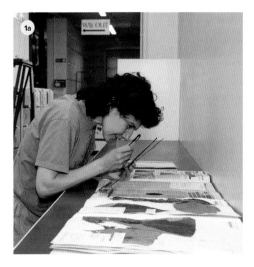

1a+b Large workspaces, suitable to spread out specimens for curation and research work. **2** Using a microscope. **3** Trolley workspace. **4** Specimen mounting unit.

Types of workspaces in the herbarium
- Sorting tables with optimal lighting
- Desk for microscope, dissections and computer work
- Specimen preparation area
- Specimen mounting unit
- Quarantine area
- Incoming and outgoing processing space

- Digitisation
- A separate room for spirit (wet) collections, poison treatment with a fume cupboard and sink
- Offices for curation, research, collections management
- Separate room for eating and drinking

HEALTH & SAFETY
RISKS IN HERBARIA
Elaine Porter

The health and safety (H&S) of both people and collections are paramount. The mantra all institutions should be working towards is: 'safe place, safe persons, safe equipment and safe practice'; all this is done by assessing risk.

Managing risks

In many countries assessing risks is a requirement under law; for example, in the UK, this falls under the Health & Safety at Work Act etc. 1974, in the USA – the Occupational Safety and Health Act of 1970, in Australia – the Work Health and Safety Act 2011. Risk assessments aim to reduce the risk to 'as low as reasonably practicable' (ALARP), providing protection for people, collections and the buildings that house them. Institutions must have a H&S policy, fire risk assessments (FRA) and related emergency plans in place, which are regularly reviewed. Additionally, robust induction and refresher training processes must be in place to ensure that everyone is fully aware of the protection and disaster protocols. All herbaria should aim to fully comply with country and local regulations and/or implement such assessments to ensure the safety of their collections, their staff and visitors.

Water damage

Water ingress or the use of water-based fire suppression can damage collections and cause issues with humidity. Basements of herbaria, particularly those that are close to rivers, flood plains or the sea may be particularly vulnerable to water ingress. The installation of water detection systems to reduce the risk of water damage is paramount. Detection systems are a cost-effective and easy way to minimise this risk to collections.

1 Water damage to cardboard/ paper boxes. **2** Water damage to paper. **3** Faulty wiring in a plug. **4** Burnt-out plug sockets caused by overloading. **5** Comparison of LED and florescent lighting (LED below/Florescent above).

Gas and chemicals

Alcohol used in the storage of specimens, e.g., spirit collections, can be a significant risk in herbaria. These spaces should be well managed, protected using temperature and humidity regulation and monitored using gas detection systems, which in turn must be regularly serviced. Chemicals should be stored in a ventilated, bunded (i.e., tank within a tank) dedicated unit which is kept secure. Spillage kits should be available; these are an emergency stock of absorbent materials used to clean up liquid spills, which can be purchased from retailers across the world.

Lighting

Natural lighting is the best, but not always practicable – so fluorescent-tube lighting is often used. Where possible, switching to LED lighting can help to reduce heat production, for which traditional fluorescent tubes are renowned. Reducing fire risk and reducing electricity consumption are key aspects for all new sustainably built buildings. For the comfort of those reading or writing in low natural daylight conditions, it is advisable to provide desk lamps to supplement LED tube lighting.

Electricity

Although many institutions have moved away from mainframe computers using 'cloud' storage options for data, the requirement for laptop and internet access has increased. Thus, to reduce the reliance on using multiple extension leads 'daisy-chained', which are a fire risk, sufficient power points are required to accommodate all required electrical equipment, e.g., laptops, microscope lights, boiling rings (where used) etc. For example, in the UK, a Portable Appliance Testing (PAT) regime (often done yearly) checks the integrity of such appliances. The need for telephone points may be less necessary if institutions use computer software for communication.

Ventilation

Adequate ventilation is essential. It is important to aim for the renewal of air, while avoiding excessive loss of pest-preventative substances or the entry of humid air or dust. Good ventilation contributes to staff comfort, while lack of ventilation can be a health hazard, even more so post the 2020 covid-19 pandemic. To further protect staff, it is important to segregate work such as that with volatile substances (e.g., spirit collections) by utilising fume cupboards.

PROS OF NATURAL VENTILATION
- Environmentally friendly
- Less expensive than mechanical ventilation

CONS OF NATURAL VENTILATION
- Very hard to control – relies on wind speed/direction
- Requires windows to open
- Open windows can increase pest/dust entry

PROS OF MECHANICAL VENTILATION
- Easier to automate and control
- Regular exchange of fresh air guaranteed if using air handling units

CONS OF MECHANICAL VENTILATION
- More expensive than natural ventilation
- Uses more electricity

Maintenance

A building, no matter how well conceived and built, is only as good as the maintenance that it is given thereafter. Therefore, all the built-in systems (fire detection, fire suppression, gas detection) must be regularly serviced. Planned preventative maintenance (PPM) should be undertaken and recorded to ensure that general maintenance such as guttering clearances, fire extinguisher servicing, etc. and electrical testing of fixed and portable appliances (PAT testing) are undertaken.

RISKS IN HERBARIA – FIRE
Elaine Porter

Worldwide, several important collection buildings have been destroyed by fire with the loss of valuable scientific material, the product of many years of work. Strict fire precautions must be enforced, including a complete ban on smoking and the use of naked flames.

Fire precautions

Fires can start naturally, e.g., forest fires (Nordling 2021), or via man-made means, e.g., electrical faults, contractor error. No matter the cause, the Fire Risk Assessment (FRA) for a building should highlight the risks and mitigations in place. All rooms, corridors and stairways must be separated by adequate compartmentation, using walls and fire doors. Regularly maintained fire extinguishers must be placed in all areas of the building. Advice must be sought, probably from the local fire brigade, concerning the installation of fire hydrants and other portable firefighting equipment.

The possible damage to the collections by water used to control fire should be considered and, where possible, alternative fire suppression systems should be invested in. A fire-alarm system must be installed, and a plan made for the procedures to be followed if a fire does occur. Fire warden/marshal sweeping protocols for evacuations can help to reduce the time taken to evacuate a building and at the same time allow for some protection of collections. All staff and visitors must be familiar with the building evacuation procedures and regular fire-drills should be held.

Fire suppression

Suppression systems should be installed into a building to help extinguish a fire as quickly as possible. Most have built-in detectors that monitor for heat, smoke, or other warning signals. These detectors are attached to an alarm system, which activates when a fire has been detected and initiates steps to release a substance, e.g., water, inert gases, etc. that suppresses the fire.

It is imperative to understand the differences between systems and select the best one for the required task, since the wrong one has the potential to cause more harm than an actual fire.

For example, water sprinkler systems douse water liberally into areas where they were installed. However, advances in technology have improved these into water mist systems, which produce ultra-fine droplets using 50–90% less water, resulting in less collateral damage. Whereas clean agent systems use environmentally friendly chemicals to extinguish fires. Often inert gases, nitrogen, argon and carbon dioxide or a combination of two or more, are used to reduce oxygen levels to a point where combustion cannot be maintained. In herbaria the use of less water for fire protection is of course desired, and a water mist system may be the way forward if cost and practicality preclude the use of other types of systems.

Water v clean agent fire suppression

Water-based fire suppression

PROS

- Relatively inexpensive to install
- Newer misting versions use less water than traditional water systems
- Water mist droplets surface area 100 x greater than sprinkler water

CONS

- Traditional systems use a lot of water = damage
- Traditional systems often need a tank
- Water or mist droplets do not reach all parts of a protected space
- Presence of water can increase humidity levels
- Clean-up can be expensive and time-consuming
- Misting systems not suitable for large areas

Clean agent fire suppression

PROS

- No water used = no water damage or humidity issues
- Gases can reach all areas of a protected space
- Environmentally friendly
- Cleaning up is cheaper and quick

CONS

- Installation cost
- Size of area to protect can affect effectiveness
- Protocols must be in place so that persons are not asphyxiated

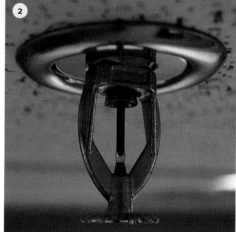

1 Example of a fire-damaged herbarium specimen.
2 Example of a sprinkler head. **3** Cannisters containing argon/nitrogen gas mix for use in extinguishing fire in collection spaces.

EMERGENCY PLANNING
Elaine Porter

Creating an emergency plan is vital to ensure that all risks have been identified and mitigations, e.g., manual and engineering solutions, are put in place. Emergency planning is threefold – plan to protect, plan to respond to an incident, and review regularly.

The plan
The emergency plan must include evacuation procedures, fire strategy, building plans (including stop cocks; fire detection, fire alarm and suppression systems), disaster and salvage plans and contingency plans. Lists of staff, as well as volunteers, who are local and live close by who could be called upon to come to help are of value. The plan is a working document and should be tested and regularly reviewed.

There are companies available who can help institutions create a plan and give training to staff. Some like Harwell Restoration (UK; see Harwell Restoration 2021), Australian Disaster Recovery, and National Heritage Responders (USA) provide services for emergencies, such as temporary offsite storage for specimens in stable conditions and specialist support, etc.

Training
The best way to test the robustness of the plan is to

a. Ensure that its contents are disseminated to all persons (e.g., staff, fire wardens/marshals, maintenance crews, cleaners, salvage volunteers) via induction sessions.

b. Provide training, so that people know what to do, where the disaster kit is kept, how to use the equipment to salvage specimens, etc.

c. Test it (which could include the local fire service) so that everyone can put their training into practice.

Stores and equipment
Herbaria should have an 'emergency store' to enable specimen protection and recovery in event of an emergency. These stores should be situated on the ground floor and with a sister kit held in a separate location. At a minimum it should contain:

- Adsorbent pads/socks
- Brooms/mops and buckets or aqua vacuums
- PPE – gloves, masks, aprons, foot covers or all-body suits
- High vis jackets
- Rubbish bags
- Blotting paper
- String, Sellotape, scissors
- Pencils and note pads
- Labels (tie-on variety)
- Torches and batteries
- Spillage kit (including granules)
- Sandbags
- Cordon tape (red/white)
- Light sticks
- Polythene sheets (which are large enough to drape over cupboards)

It may not be economically or physically practical to hold water pumps, storage boxes/crates, fans, etc. on site but it is sensible to record in the plan where these items could be acquired from at short notice in case of a disaster.

TIPS
- Have an inventory
- Regularly check the stock
- When stock is used, replace it immediately
- Never lock the emergency store
- Keep staff training up to date

1 Salvage training session, RBG Kew.
2–4 Suggested disaster stores contents.

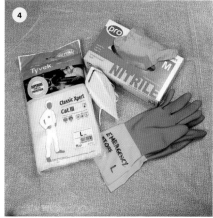

PEST MANAGEMENT
INTRODUCTION
Alison Moore

Herbarium collections are vulnerable to damage from sources such as water, fire, humidity and pests. This section will focus on the best defence against such damage – a proactive Integrated Pest Management (IPM) policy highlighting the control of the physical environment and reduction in the use of harmful chemical treatments.

Integrated Pest Management (IPM)
A successful IPM policy should be part of an institute's Collections Management policy and needs to cover:

- The collection as a whole, identifying priorities throughout
- Staff: who is responsible, what duties do they have, what training do they require?
- The building as a whole, managing all areas, including those without collection spaces
- Prevention, monitoring and emergency procedures, including regular reviews and environmental assessments
- A preventative regime is less time-consuming and costly to maintain than remedial eradication treatments

Insect biology
An IPM policy must understand the biology of the pest species. Most insects follow a similar life-cycle: adults lay eggs, which transform into larval or nymph stages, larvae form pupae while nymphs grow and moult, finally emerging as new adults. It is while larvae or nymphs that most of the feeding occurs, and hence, this is usually the most destructive stage of development. Stages of the life-cycle can look markedly different from each other, so staff should be trained to recognise the appearance of all stages in order to spot them during both monitoring and general herbarium work.

Pest species have specific environmental conditions for optimal reproduction and development times (see Pinniger 2015); in general, temperatures of 20°C and above encourage insect breeding; higher relative humidity and temperatures will speed up the life-cycle and improve the success of the pest species.

Identification and awareness
Correct identification of the pest species is paramount, and the source of infestation must be discovered. This involves identifying signs of infestations, i.e., frass, casings, webbings, skeletal leaves, or holes in collections/furniture, as well as the pest itself (eggs are generally minute and not useful for identification). Furthermore, it is necessary to be aware of pest species present among the local fauna, and also those emerging as pests in herbaria worldwide, thus presenting a future risk.

FURTHER READING
IPM – Working Group Identification Resources. https://museumpests.net/identification/identification-resources/

1 Staff training – pest identification course. 2 Reference collection. 3 Berlin beetle larvae damage to specimen. 4 Carpet beetle larvae damage to woollen door seals. 5 Biscuit beetle damage to raffia mat. 6 Mouse damage to specimen sheets and flimsies. 7 Mould damage to a Myristicaceae specimen. 8 Museum nuisance beetle damage to cabinet.

TIP

Rogues Gallery

It can be useful for identification to have a reference collection of pests encountered which have been verified by an expert (see Fig. 2).

How many is too many?

Limited numbers of some pests, e.g., booklice, silverfish and even cockroaches, can be expected and tolerated within buildings, but even a single individual of others cannot be, for the safety of the collections. For such destructive pests as biscuit, tobacco and Berlin beetles, eradication must be immediate and complete to prevent further spread and damage to the collection.

COMMON PESTS – INSECTS
Alison Moore

Beetles and moths are the worst herbarium pests in cool, temperate regions of the world. In warm temperate and tropical climates, termites, cockroaches and woodborers all cause major problems while booklice, woodlice and silverfish dominate in humid climates. The most common pests have more cosmopolitan distributions.

In the following sections, the most common pests are discussed, and examples of life stages and damage caused are illustrated.

Cellulose feeders

Evidence: casings; neat, round holes where adults emerge; flower and fruit damage.

- Biscuit/drugstore beetle – *Stegobium paniceum* cosmopolitan, adults active fliers; 2–3 mm, distinguished by rows of pits on elytra and enlarged final three segments of antennae. Larvae, C-shaped, feed for 5–16 weeks or more on herbarium and bulky material
- Cigarette/tobacco beetle – *Lasioderma serricorne* widespread tropical/subtropical, in heated buildings in Europe, adults strong fliers; 2–3 mm, distinguished by saw-toothed antennae, larvae (almost identical to *Stegobium* larvae) feed for 5–10 weeks or more on herbarium specimens, especially seed heads and labels
- Berlin/Stockholm beetle – *Trogoderma angustum* S. American origin, now regularly encountered in Europe/Scandinavia, outside of the tropics, needs a heated building to complete development. Adults short-lived, 2–4 mm; distinguished by three bands of white hairs on elytra. Larvae similar to *Anthrenus* with longer tufts of hair at tip, feed on dried plant material, especially flower heads

Keratin feeders

Evidence: matted organic debris and damage trails/holes, cast (hairy) casings. Distinctive adults (beetles).

- Carpet/fur beetles – *Anthrenus, Attagenus* and *Anthrenocerus* cosmopolitan, adults active fliers, 2–3 mm distinguished by pattern of scales on elytra, will feed on pollen (especially Asteraceae). Densely hairy larvae of *Anthrenus* unmistakable 'woolly bears' initially tiny, very mobile, can feed for up to 36 months growing to 5 mm long, will attack woollen felt strips in cabinets. *Attagenus* larvae similar but longer (< 8 mm), will eat seeds
- Webbing clothes moths – *Tineola bisselliella*, case-bearing clothes moth – *Tinea pellionella* cosmopolitan, highly problematic in heritage collections. Adults 10 mm, pale moth, does not feed. Larvae spin silk webbing, which is used to pupate within or graze under respectively. Damage woollen felt strips in cabinets
- Others: Indian meal moth – *Plodia interpunctella*, brown house moth – *Hoffmannophila pseudospretella*, white-shouldered house moth – *Endrosis sarcitrella* all more or less common in museums worldwide

- Museum nuisance/American wasp beetle – *Reesa vespulae* N. Europe, USA, Scandinavia. Adult 2–4 mm with distinctive band of yellowish hair on elytra. Larvae, up to 6 mm, tubular and hairy, feed on insect collections but will attack seeds. Parthenogenic

- Odd beetle – *Thylodrias contractus* Central Asian in origin, important pest in USA, Scandinavia, rarer elsewhere in Europe. Adult rarely seen. Larvae; 2–3 mm with bands of short, pale hairs

- Hide beetles/leather beetles – *Dermestes* sp. cosmopolitan. Sometimes used in museums to clean carcasses but may escape and persist in unhygienic areas. Adults 7–9 mm, antennae end in 3-segemented club. Larvae, 'fuzzy' up to 12 mm, bore holes in wood before pupating, thereby damaging cabinets

1 *Stegobium paniceum* (biscuit beetle). **2** *Stegobium paniceum* (biscuit beetle) larvae. **3** *Lasioderma serricorne* (cigarette beetle). **4** *Trogoderma angustum* (Berlin beetle). **5** *Trogoderma angustum* (Berlin beetle) larvae. **6** *Anthrenus verbasci* (varied carpet beetle). **7** *Anthrenus verbasci* (varied carpet beetle) larvae. **8** *Attagenus smirnovi* (vodka beetle). **9** *Attagenus smirnovi* (vodka beetle) larvae. **10** *Tinea pellionella* (case-bearing clothes moth). **11** *Reesa vespulae* (museum nuisance beetle). **12** *Reesa vespulae* (museum nuisance beetle) larvae. **13** *Thylodrias contractus* (odd beetle) larvae. **14** *Dermestes peruvianus* (Peruvian larder beetle). **15** *Dermestes* larvae.

OTHER PESTS
Alison Moore

Insects – detritus and mould feeders, scavengers

More of a nuisance until populations gain significant numbers, usually indicators of unfavourable environmental conditions, e.g., damp, high relative humidity, unhygienic conditions. All the following have a cosmopolitan distribution.

Evidence: sightings of adults

- Spider beetles – *Ptinus* sp., *Niptus hololeucus*, *Gibbium psylloides*, more common in temperate climates. Omnivorous, will attack herbarium specimens
- Cockroaches, especially *Blattella germanica/Blatta orientalis* (German/Oriental cockroach), *Periplaneta americana* (American cockroach). Tropical, also inhabit buildings in temperate climates, omnivorous, carry allergens and can trigger asthma, interference with electrical equipment creates a fire risk
- Silverfish – *Lepisma, Ctenolepisma*, will damage paper/cardboard and eat labels of specimens to get at glue underneath
- Plaster/fungus beetles including Latridiiae, Cryptophagidae, Mycetophagidae, feed on mould; indicators of persistent damp with associated fungal growth
- Booklice, Psocids, especially *Liposcelis bostrychophila*. Abundant in humid climates and buildings with high temperatures. Feed on microscopic fungi, mounting paper, labels, pollen and delicate flowers. Parthenogenic

Insects – woodborers

Adults do not eat but larvae very destructive. Develop in living timber, cut or decaying wood, often in association with or dependent on wood-decaying fungi. Full development of e.g., *Anobium*, *Xestobium* indicates higher than optimal humidity.

Evidence: emergence holes in wood and paper pulp. Larvae take significant time to emerge hence damage often overlooked

Examples include: furniture beetle/woodworm – *Anobium punctatum* Cosmopolitan. Death watch beetle – *Xestobium rufovillosum* Europe, N. America. Longhorn beetles – Cerambycidae cosmopolitan, increasingly a pest of buildings. Bostrychid borers, tropical. Powder post beetles – *Lyctus* sp. Tropical, increasingly common, distinctive floury dust and frass from larval tunnels. Termites – Isoptera, tropical and warm temperate climates. Subterranean termites may become more of a problem in temperate climates as a result of climate change.

Other animals

Mammals (e.g., rats, squirrels, possums) cause damage to the structure of buildings, electric and ventilation systems. Bird (especially feral pigeons) nests are a source of insect pests. Geckos are insectivores – a high density can indicate an insect problem.

Droppings of vertebrates are unsightly, often corrosive and carry diseases. Carcasses become a protein source for other pests. Many vertebrate pests are nocturnal; evidence found during the day includes droppings, holes, gnawed food/packaging, footprints, smears (from rodents' regular walking routes).

Woodlice – Isopoda indicate persistent damp and high humidity.

Other important pests found in or near collections. **1** *Niptus hololeucus* (golden spider beetle). **2** *Adistemia watsoni* (plaster beetle). **3** *Lepisma saccharinum* (silverfish). **4** *Liposcelis bostrychophila* (common booklouse). **5** *Reticulitermes lucifugus* (subterranean termite) worker. **6** *Pecellio scaber* (wood-louse). **7** *Blatta orientalis* (Oriental cockroach). **8** *Anobium punctatum* (common furniture beetle).

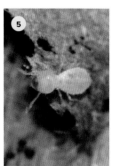

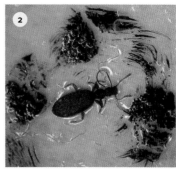

Taxa most at risk from insect attack

- Families with a high starch content: Apiaceae; Compositae; Brassicaceae; Convolvulaceae; Ericaceae; Leguminosae; Lamiaceae; Liliaceae (s.l.); Orchidaceae; Orobanchaceae; Rosaceae; Salicaceae; Scrophulariaceae; Solanaceae

- Plants which trap insects: insectivorous plants, e.g., Nepenthaceae, or those with intricate pollination mechanisms, e.g., Apocynaceae-Asclepiadoideae

What a pest needs

- Shelter – entry routes, undisturbed dead spaces, lax housekeeping
- Warmth – insects breed readily at temperatures over 20°C
- Moisture – from the environment or in food sources, ambient air relative humidity greater than 60% will increase problems with some pests and mould growth
- Food – dried plant material a ready source of starch, human food and waste attracts rodents, larvae fond of most animal or vegetable material

AVOIDING AND PREVENTING PEST INFESTATIONS
Elizabeth Woodgyer

A successful strategy for preventing pest infestations is dependent on keeping collection areas clean, well maintained, cool and dry; making the environment inhospitable to pests by removing places for them to live and breed; and decontaminating all incoming dried plant material to prevent pests from other sources from being introduced into the collections.

1 A cluttered work area hampers cleaning. **2** Unprotected herbarium sheets. **3** Unprotected specimen bundles. **4** Keep plants and shrubs away from outside walls. **5** Underfloor ducts trap dirt and dead insects. **6** Door with bristle strip.

Checklist of pest prevention measures

- A thorough, regular cleaning regime including vacuuming (preferably), sweeping and dusting: debris, rubbish and dead insects must not be allowed to accumulate

- Keep work areas tidy, avoid clutter which hampers effective cleaning

- Avoid carpeting in collection areas and offices – carpets hinder pest detection, collect detritus and are more difficult to clean than smooth floor coverings

- Store specimens within cabinets, drawers, boxes or sealed polythene bags and avoid leaving unprotected specimens out on desks, particularly near windows

- Close herbarium cabinet doors firmly and keep in good repair with insect-proof seals to doors

- Aim to keep collection areas cool at a temperature of 16–18°C and at a relative humidity of 40–60% (see Environmental Conditions)

- Measure and monitor temperature and humidity levels in collection areas using instruments, e.g., thermometers, hygrometers, data loggers

- Do not allow food and drink in collections areas or offices and clearly designate a separate area for food consumption, storage and disposal with food items to be stored securely in airtight plastic containers

- Do not allow live plants or cut flowers within the herbarium building as they can harbour insects, which can provide food for herbarium pests

- If feasible, nominate a pest management coordinator with responsibility for pest prevention and control, but all herbarium staff should be aware of pests and IPM

Keeping pests out of herbaria

Well-designed, modern, air-conditioned buildings can successfully limit the entry of pests, but excluding them from old, historic buildings is more challenging.

Minimise access by:

- Good maintenance of building infrastructure: well-fitting windows, brickwork and ventilation grills kept in good condition, well-sealed openings etc

- Preventing damp e.g., repair leaks from gutters, water pipes or roofs to prevent water ingress and any resultant mould and fungal damage

- Blocking unused chimneys and fireplaces after deep cleaning

- Preventing birds and mammals from nesting or roosting in/on the building through use of specially designed products e.g., wire mesh, netting, spikes and wires

- Keeping live plants and shrubs away from outside walls, particularly near windows

- Fitting doors and windows with draft excluders and bristle strips

- Keeping windows and doors closed wherever possible. If they must be kept open for ventilation, they should be fitted with fine fly mesh screens

- Cleaning heating pipes and ventilation ducts regularly

QUARANTINE AND DECONTAMINATION PROCEDURES
Elizabeth Woodgyer

It is essential pests are not introduced to the collection, therefore, all dried plant material must be decontaminated as soon as possible before entering the herbarium, ideally in a separate unpacking room or building, situated near the parcel delivery area.

How pests may be introduced
- Newly acquired material from fieldwork, other institutions or private homes
- Loans sent or returning from other institutions
- Specimens or objects brought in from other departments
- Visitor's own material for study

Deep freezing
- A safe, effective and practical method of decontamination. The freezing process will kill insect adults, larvae, pupae and eggs of temperate and tropical pests alike

- Top opening, domestic chest freezers which require manual defrosting are most suitable for small herbaria. Large herbaria often use commercial walk-in freezers, which can accommodate many specimens at a time and large, bulky items

- Any protective packing material must be removed prior to freezing or it will act as insulation and prevent the cold from penetrating into the specimens

- NOTE – packing materials can harbour insect pests and must be frozen if they are to be reused

- Items and bundles must be sealed inside heavy-duty polythene bags to prevent damage by humidity changes or moisture migration. Excess air in the bag should be removed before sealing (Pinniger 2015)

- Specimens should be treated for: 72 hours at −30°C, 7 days at −25°C, or 14 days at −18°C (Pinniger 2015). Extra chilling time should be added to allow the freezer unit to reach the critical temperature after loading and for the specimen bundles to reach the required temperature at their centre

- The freezing cycle should run uninterrupted and after freezing, material should remain in the polythene bags until it has returned to room temperature to prevent condensation damage

Material unsuitable for freezing
Some ethnobotanical objects, e.g., lacquer, resin and rubber, or items too large for the freezer, cannot be frozen but instead should be isolated/quarantined by sealing inside clear polythene bags and waiting to see if adult insects emerge from the object (Timbrook 2014).

MONITORING AND RECORDING PESTS
Elizabeth Woodgyer

Users of the herbarium collection should always be alert for signs of insect activity such as frass, shed casings, damaged specimens, holes in covers and alive or dead adults or larvae. Regular monitoring and recording are key to catching an infestation early before a serious outbreak can occur and irreversible damage is done (Pinniger 2008). It also provides valuable information on the distribution of insects in the building, identifies high-risk areas and pinpoints environmental sources of infestation.

Trapping insect pests
Traps enable insects to be detected at very early stages of infestation, greatly improving the chances of preventing damage (Pinniger 2015). They should be set wherever collections are worked on or stored. After surveying the building, prepare a plan and place the traps in a regular grid pattern if possible. Only set as many traps as is manageable. Date and number each trap and mark its position on the plan. This ensures consistency when comparing results over time.

Traps should be checked at regular intervals, preferably every three months. They should be replaced annually, or more frequently if they become dirty, lose their stickiness or become covered in non-pest insects, which could become food for pests.

Types of insect trap
Most effective are sticky traps, which can either be used to reveal an existing pest infestation, or to monitor for new problems (Pinniger & Lauder 2018).

Blunder traps
- Non-toxic, non-specific, cheap and safe to use. They have a sticky adhesive interior which indiscriminately traps crawling insects and larvae

- Traps should be placed on the floor in the angle against the wall, where they are unlikely to be moved or trodden on and not in open areas
- Each trap should display a clear DO NOT TOUCH warning

Pheromone traps
- Pheromones of certain insects are available as synthetic lures. Before investing in expensive lures, it is vital to correctly identify the pest insect
- The most successful lure is for webbing clothes moth (*Tineola bisselliella*); others are available for cigarette beetle (*Lasioderma serricorne*), biscuit beetle (*Stegobium paniceum*) and carpet beetles (*Anthrenus* spp.) (Pinniger 2015)
- Follow the manufacturer's advice regarding the placement of pheromone traps

Electric traps
- A low-wattage light bulb attracts insects onto a spherical adhesive disc. They are marketed for pet fleas but are effective in attracting other pests such as biscuit beetle (*Stegobium paniceum*) and Berlin beetle (*Trogoderma angustum*)
- They should be positioned on a bench or table and kept switched on at all times

Recording results

Insects caught on the traps are identified, counted and recorded. Correct identification and recording of life stage are essential. A database should be used to collate the data and monitor the situation over time – simple versions are available for free on the internet, or a basic spreadsheet can be created. This should hold information on trap locations, date of checking and what was found in terms of insect types and numbers (Museum of London 2013). A downloadable Insect ID Monitoring Sheet is available from the English Heritage website (www.english-heritage.org.uk). The results from the inspections will give a good indication if there is a potential problem and highlight where any preventative measures or remedial treatments need to be focused.

Interpretation of catches

• Identification indicates whether an insect is a pest or harmless
• Identification of pest types indicates what material is at risk
• Monitoring over time indicates whether an infestation has become established or is an isolated outbreak
• Larvae indicate that the species is breeding within the building
• Silverfish, booklice and plaster beetles can suggest local high humidity and damp conditions
• Woodlice and centipedes suggest poor sealing around doors and windows
• Presence of dirt and fluff can indicate poor cleaning and general maintenance (Child 2011)

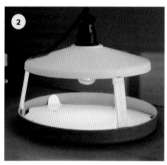

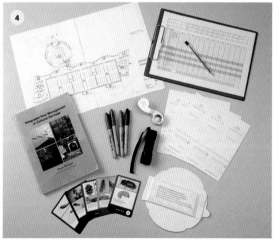

1 Blunder trap in position at wall/floor angle. **2** Electric insect trap: the heat and light emitted attracts insect pests. **3** Blunder trap, checked and trapped insects recorded. **4** Monitoring equipment should include: Hand lens (loupe), ideally with an LED light, for checking and identifying; torch/flashlight; plan of the building; clip board and monitoring sheet; coloured permanent marker pens to circle catches and ensure they are not counted in subsequent surveys; IPM book/images to aid identification; spare traps.

CONTROL OF INFESTATIONS 1. PHYSICAL
Alison Moore

After pest identification, the source of the infestation must be found, remedied and physical problems fixed (e.g., water ingress prevented, damage fixed, nests removed). The site must be thoroughly cleaned, removing all detritus, frass and dead insects. The infestation must then be treated to prevent further damage and reinfestation.

1 Boxed specimens in Integrated Contamination Management (ICM®) chamber (heat treatment). **2** Pheromone lure with trapped *Tineola bisselliella* (webbing clothes moth). **3** Pheromone lure for *Lasioderma serricorne* (cigarette beetle).

Physical control

The safest methods of control which are concurrent with a preventative approach.

Temperature control

Freezing rapidly reduces the metabolism of the insects until they are killed. Protocols should be followed as for preventative freezing (see page 202).

Heating: Treatment at 52°C will kill most insects in 1–2 hours (Child 1994), allowing up to 18 hours for this temperature to reach the centre of the material and controlling relative humidity to ensure collections are not damaged by drying out. Portable environmental chambers with heaters, or sealed bags within drying chambers can also be used (IPM-Working Group, Solutions Subgroup 2018). In hot climates, solar heat can create a microclimate around specimens wrapped in black plastic surrounded by clear plastic and left, fully exposed in the sun, for a number of hours (depending on the density of material to be treated; for treatment guide and estimated times see Strang & Kigawa 2009).

Anoxic treatments: Oxygen levels below 0.4% will kill insects at temperatures over 20°C due to oxygen deprivation and desiccation. The process requires a gas-tight chamber or tent, sensors for oxygen, temperature and humidity, and a gas source (usually nitrogen). Low oxygen levels can then be maintained using oxygen scavengers (substances which react with oxygen, reducing its concentration) for a period of up to 8 weeks depending on species. Expensive, specialised chambers are available for collections (Maekawa & Elert 2003). This can be a very useful method for fragile and precious specimens susceptible to damage by other methods.

Biological control

- These techniques are restricted to collections housed at temperatures above 15°C with very low levels of biocide contamination (Frick & Greeff 2021). International regulations on the use of introduced organisms differ and should be consulted prior to use

- Pheromone lures used as part of monitoring programmes (see page 203), can reduce fecundity by mass trapping, and pinpoint source of infestations

- Parasitoids develop on or within insect hosts at a particular part of their life-cycle and can suppress pest population numbers. Examples include parasitic wasps *Lariophagus distinguendus* to control biscuit and tobacco beetles (Pinniger 2015) and *Trichogramma evanescens* to control webbing and case-bearing clothes moths in historic houses (nationaltrust.org.uk)

CONTROL OF INFESTATIONS 2. CHEMICAL
Alison Moore

A successful Integrated Pest Management (IPM) programme will always seek to minimise, if not eliminate, the need for the application of chemical pesticides.

Chemical control
In many parts of the world, wholesale use of chemicals to treat and prevent infestations is becoming less common. Changes in legislation and further evidence of serious environmental impact and health hazards means that previously common practices are now banned or discouraged. Given the changing nature of chemical legislation, it is not of value here to list current chemicals in use. Therefore, a broader overview of treatments follows. Whichever methods are employed, it is vital that local, national and wider regulations are consulted before any regime is started, all manufacturers' recommendations are followed and staff are not subjected to potentially harmful chemicals (see next page for regulatory bodies).

Fumigation refers to the use of gaseous pesticides to eliminate insects from an environment by poisoning or suffocation. Ideally, procedures are carried out in specially built chambers or under gas-proof sheets, but buildings may be made safe by using specialised sealants on exit points. The gas must remain long enough to penetrate infested materials and kill all insects (if less time is spent, as an economy consideration, insect resistance may develop, thus escalating the problem). It then needs to be aired off before the area under control can be safely re-entered. Sulphuryl fluoride is one of the few chemicals for use in this process that is not now routinely banned. Fumigating with 60% carbon dioxide is effective but must be carried out by a specialist.

Space sprays are used against flying or crawling insects by way of smoke or fogs of (usually) pyrethroid-containing mixes.

Residual sprays allow empty building or storage areas to be treated. Concentrated liquid or powdered pesticides diluted in water or ready-to-use micro-emulsions can be applied with a pressurised hand sprayer. Alternatively, pesticide dusts can be applied with a hand-duster or aerosol formulation to treat voids, cracks and crevices.

Insect growth regulators and chitin inhibitors disrupt hormones involved in life-cycles. Strict licensing laws exist in many areas but can be effective against termites and cockroaches.

For further information on the range and types of pesticides see Pinniger (2015).

Trapping non-insect pests and baiting
Aside from trapping as part of monitoring for insects, live capture or snap traps can be laid when rodent activity is detected. Rodenticide is most widely applied as edible baits; highly toxic to target species as well as humans and other animals so great care is needed. Correct placement of traps and bait stations is needed to bring infestations under control and regular monitoring will assess the success or otherwise of the technique. Old baits may support insect infestations and must be regularly removed. In most cases, specialists in rodent control should be contracted.

Legislation

Many regulatory bodies worldwide control the use of pesticides and biocides, and most countries set out guidelines for those approved for use in collections under specific application and safety precautions.

Under the International Treaty of the World Trade Organization (WTO), the *Agreement on the Application of Sanitary and Phytosanitary Measures* holds all countries to account for the use of products harmful to human health.

For European, Chinese and American regulation:

Europe: check with the Biocidal Product Regulation, part of the European Chemicals Agency (ECHA) legislation https://echa.europa.eu/legislation

China: https://www.cirs-reach.com

US: https://www.epa.gov/pesticides

In countries with no specific regulations: information and guidance from the ECHA is often used in implementation.

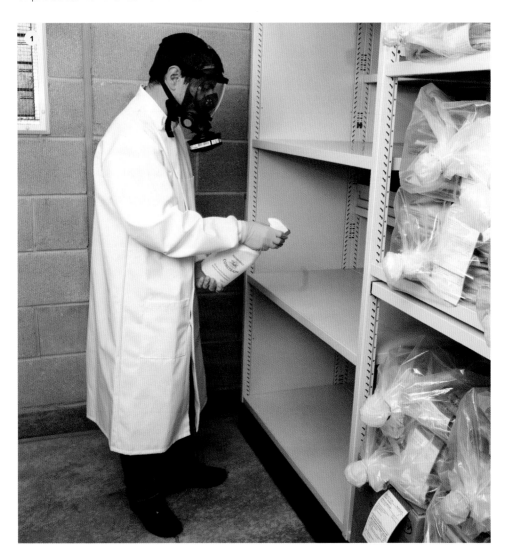

Mould

Mould will develop in herbaria when relative humidity levels are high (> 60%) and air circulation is poor. Fungal infestations develop quickly and cause visible damage to specimens and mounting paper, detectable by eye or hand lens, from tiny, stain-like discolorations to large furry areas of growth.

(Moulds which appear dry can be removed with a special hazardous material vacuum cleaner (equipped with an Ultra-Low Particulate Air (ULPA) filter)) or brushed off gently and the spores vacuumed immediately. Moulds which appear wet can be bagged and frozen to prevent further growth, or specimens should be cleaned with diluted ethanol or isopropanol (7 parts plus 3 parts water). Storage areas need to be cleaned in the same way with a microfibre cloth, making sure all materials are disposed of or decontaminated using biosecure measures.

Fungal hyphae can penetrate and inhabit specimens, subsequently becoming a food source for insect pests while numerous spores may be released by fungi into the environment, contaminating rooms, objects and air conditioning systems. Infested specimens are very vulnerable to reinfestation and should be monitored (Frick & Greeff 2021).

Any residual stains should be treated by a conservator.

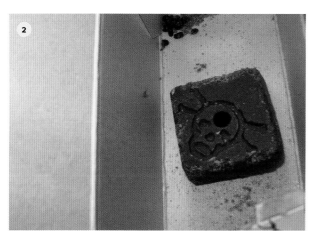

1 Trained staff member with safety equipment for residual spraying.
2 Rodent bait.

HERBARIUM HIGHLIGHT:
HERBARIUM BOGORIENSE (BO)

Himmah Rustiami, Marlina Ardiyani and Ina Erlinawati

Herbarium Bogoriense (BO) is one of many historical herbaria in Asia but is the oldest herbarium in South-East Asia being founded on 14 August 1841. With almost one million herbarium specimens, BO is an important source of biodiversity information for the entire South-East Asia region, and especially Indonesia.

Herbarium Bogoriense buildings through time. **1** First building of Herbarium Bogoriense inside Bogor Botanic Gardens. **2** Juanda Street premises, 1960–2007. **3** Herbarium Bogoriense at the Cibinong Science Center. **4a–c** The Collections at BO, showing the modern cabinets with internal boxes.

Collectors and collections

The Herbarium Bogoriense (BO) houses an estimated 960,000 herbarium specimens (based on 2021 data), including 38,100 specimens in spirit, 6677 carpological collections, and 17,494 type specimens which are already digitised with around 7000 type specimens available online. The collection is increasing by 2000–3000 specimens per year, through field expeditions by BO staff, as well as gifts and exchanges with other institutions in Indonesia and overseas. The Herbarium consists of specimens from all major islands in the Malesian region, namely Borneo, Sulawesi (Celebes), Java, Lesser Sunda, Moluccas, New Guinea, Philippines and Sumatra, as well as Peninsular Malaysia and southern Thailand; BO also holds collections from other areas outside Malesia, such as mainland Asia.

Buildings

The first BO building was inside Bogor Botanic Gardens (Fig. 1), before it was moved in 1960 to Juanda Street, just across the road from Bogor Botanic Gardens (Fig. 2). In May 2007, Herbarium Bogoriense was moved into a new building in Cibinong Science Center, an area of around 160 ha, with new advanced facilities compared with the previous building in Bogor (Fig. 3).

Important collections

BO has valuable collections of Teijsmann, including type specimens described by Miquel (the director of the Rijksherbarium from 1862) in 'Flora Nederlands Indie' and 'Annales Musei Botanici Lugduno-Batavi' (Miquel 1855–1858; Miquel 1867). Subsequently, the collection increased significantly under Melchior Treub's skilful management at Bogor Botanic Gardens in the early 20th century, especially with the extensive collections of C.A. Backer from numerous locations throughout Java. Additional significant historical collections held in BO include Blume, Hasskarl, Koorders, Lorzing, Beccari, Holttum, J.J. Smith and Kostermans (see van Steenis-Kruseman & van Steenis 1950).

Herbarium management

The collections are arranged in alphabetical order, stored in the grey cabinets by family, region, genus and species (Djarwaningsih *et al.* 2002; Fig. 4). This user-friendly arrangement is easy to access, especially for finding taxa of interest. Since 1995, BO has no longer used mercuric chloride to preserve specimens, but uses freezing as the main method of pest control as well as fumigation using hydrogen phosphide, the next line of chemical defence for the specimens. Databasing is ongoing, with Magnoliopsida type collections of BO available on the GBIF website since September 2022.

Curator

Curators of BO have diverse specialisms, including fungi, liverworts, mosses, ferns and allies, gymnosperms and flowering plants, and are engaged on several projects such as updating the Indonesian Flora, and collaborating on taxonomic research with other herbaria overseas. Each taxonomist of BO has a responsibility to curate collections based on their taxa of interest.

Global Knowledge Bank

Growing gardening collections

This building holds a huge gardening treasure trove of dried plants and invertebrates from Britain's gardens, built up over 200 years.

Today these collections make up the RHS Global Knowledge Bank and hold the story of gardens in Britain. Together with the RHS Libraries, they are the world's most complete historical record of gardening, cultivated plants and the wildlife they support. Our scientists use them every day to find out more about how gardens help our wellbeing and environment, and how we can best garden in a changing world.

GET UP &
GROW

Herbarium

THE HERBARIUM IN THE WIDER CONTEXT

VISITORS
INTRODUCTION
Carmen Puglisi and Alan Paton

Herbarium visits should be beneficial for both the visitor and the institution. Herbaria host visitors with a broad range of aims and objectives for exploring the collections. The importance of these visits is outlined along with best practice for managing visitors, before, during and after the visit.

Users and uses of herbaria
The range of visitors coming to the herbarium will depend on the nature of the institution. A university herbarium may wish to encourage more student use, whereas a national herbarium may cater for a broader scientific and public audience. Carine *et al.* (2018) classified visits into 12 categories based on experience of the Herbarium of the Natural History Museum, London. These categories covered identification, taxonomy and systematics, ecology and conservation, history of science, art and literature, education, and outreach.

Benefits of visitors
Visitors can add great value to the collection by annotating the specimens, providing updated identifications and advising the researchers and curators of the host institution on particular groups, regions or collectors. Hosting visitors can help develop a better public understanding of the herbarium work, its importance in the study of plant diversity, and its relevance to environmental challenges. Where possible, access to herbarium collections should be as open as possible. However, the herbarium is a valuable and fragile resource and visitors need to understand how to work with the collection to protect it for future generations.

Engaging with visitors
An herbarium may wish to stimulate more visits from particular sectors, for example, local natural history societies, or engaging in public outreach activities such as by holding open days or producing displays explaining the work of the herbarium. If there is a high demand for identification services, a reference herbarium of typical examples of the local flora could be assembled to provide a quick way for visitors to identify their specimens or photographs.

Whatever the spectrum of users, there needs to be clear guidance on how visits should be arranged and the terms and conditions of access. The herbarium needs to have processes in place.

1 Visits focused on public engagement can be supported by the preparation of displays explaining the work of the herbarium. **2** *Quick Guide Herbarium* at the National Herbarium, SANBI (PRE) with Mr Mduduzi Nkwanyana. Representative specimens of the national flora are placed in Ziplock bags as protection from handling, and used as a reference herbarium for visitors.

BEFORE THE VISIT
Carmen Puglisi and Alan Paton

Planning is key to being prepared for successful herbarium visits. These steps include negotiating the terms and conditions of the visit and allocating the resources needed for the visitor and their work.

Website
Prospective visitors, especially if new to an herbarium, will approach the institution either through the website or by emailing the curator or a member of staff.

Herbaria should ensure that their entry on *Index Herbariorum* (Thiers, continuously updated) is as complete as possible and regularly updated. The institutional website should contain basic details, including a description of the collection, its geographic and taxonomic strengths, current research, and how to arrange a visit. There should be a contact for correspondence and/or an automated system to log visit requests.

Negotiating terms and conditions
It may be helpful to ask prospective visitors to complete an application form including personal details, affiliation, experience of work in an herbarium setting, length and scope of their visit, the material they would like to consult, access requirements, and specialist equipment needed. It may be reasonable also to fix a notice period between application and visit so that the host institution can ensure

that suitable resources, staff and staff time are allocated to the visitor. Conditions that should be negotiated in advance include any bench fees, sampling from specimens, and imaging, which will be subject to institutional policies. Visitors bringing their own material to the herbarium will need to be made aware of the procedures in place regarding pest management.

Useful documents
Once the visit is approved, the visitor should be sent information on who will act as their point of contact, normal opening times, location and transport details. Suggestions for accommodation on site or nearby should be offered if relevant. Visitors may also require a letter of invitation to obtain funding for the trip or a visa.

The herbarium can prepare in advance a series of documents for their visitors. We recommend producing a printed guide to the herbarium, a map outlining the collection arrangement of the herbarium, instructions and rules on handling and retrieving specimens, annotating, sampling and imaging material, and the institutional loan policy.

2a

Smoking

 Smoking is not permitted (including e-cigarettes) anywhere in the building, including balconies. There is a designated smoking area to the rear of the building.

Flood prevention

 Immediately report any leaks or blockages to our maintenance team on 3109 (internal) or 020 8332 3109 (external).

Pest prevention

 You are prohibited from bringing plants, cut flowers or foliage into the building due to the risk of introducing pests. Please report any pest sightings to your host who will then report to the Herbarium Pest Control Team.

Eating and drinking

 Eating and drinking is restricted throughout the building to prevent pests. Please only consume food in specified areas.

Handling the Collections

 Always wash your hands after handling specimens, and in particular before eating, drinking or smoking, as they may have been treated with pesticides. You must arrange an induction with the Collection Manager if you intend to work in the spirit collection in the basement of Wing D.

Security

 Security staff are contactable 24 hours a day. For general enquiries, contact the Constabulary on 5121 (internal) or 020 8322 5121 (external). In an emergency or to report anything suspicious call 3333 or 020 8332 3333.

Follow these security precautions:
· Don't prop open access-controlled doors.
· Don't let anyone without a staff or visitor pass through an access-controlled door behind you.
· Politely challenge anyone without a pass who is not escorted.

Restricted areas

 There are certain areas within the building where access is restricted to authorised people only. Safety signs, notices and instructions are provided to protect you – we require that you comply with them. Please ask your host if you are unsure.

If you will be visiting for more than four weeks, please ask your line manager or supervisor to organise a full induction with the Science Health & Safety Manager.

Royal Botanic Gardens Kew

Visiting the HLAA

2b

Welcome to the Herbarium, Library, Art & Archives (HLAA) building

Our Grade II listed Herbarium building houses over 7 million scientifically important specimens, making it one of the largest and most diverse botanical collections in the world. The specimens have been collected across the globe over the past 170 years, and many are irreplaceable.

Please help us keep our collection safe by taking a couple of minutes to read this essential information.

Visitors

 All visitors must sign in and out of the building at reception. You must be escorted by your host at all times. Please ensure you visibly display your visitor pass.

Emergencies

 Dial 3333 (from any internal phone) or 020 8332 3333 (externally) for help in an emergency. This number is answered 24 hours a day. Don't dial 999, as the call needs to be co-ordinated by our on-site Constabulary.

First aid

 There is a first aid room on the ground floor of Wing D. First aid kits are kept throughout the building, and each has a list of current first aiders on it.

Accidents and incidents

 Should you experience an accident or incident during your visit, report it immediately to your host, who will complete an accident report form (held at reception and with first aid boxes). Report near-misses so accidents can be prevented.

Fire

 If you discover a fire, activate the fire alarm by using one of the red break-glass call points (found in most corridors and stairwells). Phone 3333 (internal) or 020 8332 3333 (external) to report the fire.

On hearing the alarm, go directly to the nearest emergency exit and report to the fire assembly point, which is the HLAA car park (at the rear of the building). Don't attempt to fight any fires unless you have been trained to do so.

Fire prevention

 Follow these rules to reduce the risk of fire:
· Don't overload or 'daisy chain' extension leads.
· Don't prop open fire doors.
· Don't accumulate excess materials or equipment in offices or corridors.
· Don't leave items such as laptops or phones charging when unattended.
· Turn off electrical equipment overnight. Wherever possible switch off at the socket.
· All electrical items must be PAT tested before use.
· Follow all safety instructions when handling flammable items.

1 Index Herbariorum homepage.
2a+b Printed guide for herbarium visitors.

DURING THE VISIT
Carmen Puglisi and Alan Paton

Hosting a visitor requires providing support during their time in the herbarium. Aspects to consider include access to the building, health and safety regulations and supplying stationery that conforms to the curation standards in place.

Meeting the visitor
The visitor and the point of contact should agree on a mutually convenient time of arrival at the herbarium. This is so that the visitor can be shown around the building and given the essential information, including the location of emergency exits, health and safety procedures, general schedule for meal breaks, opportunities to meet with staff, access to the internet, etc. The point of contact may need to provide security clearance and a pass or key to access the building and should remind the visitor of the opening hours. During the induction it is usual to take the visitor's contact details to be used in case of emergency, in accordance with local restrictions regarding data protection.

Induction to the herbarium
At the beginning of the visit, the visitor should be introduced to the collections to become familiar with their location and arrangement. Less experienced herbarium users should be shown how to handle herbarium specimens, folders, boxes, jars and step ladders, as relevant. Health risks associated with work in the herbarium (e.g., exposure to noxious substances from preservatives in spirit collections) should be highlighted, together with any mitigating measures. It might be necessary to also discuss whether the visitor will have unassisted access to the main collection and any separate collections (e.g., type specimens, carpological, spirit).

Equipment
Different types of visitors will have different needs when working in an herbarium. Their requests might include lights, camera stands, power cables, adapters, books, hotplates and dissection kits. The host institution may want to specify ahead of the visit what material it can provide.

Researchers might be able to curate the collection by naming specimens or updating the classification of a group. To this end, they should be equipped with appropriate, archival-quality stationery and curation supplies. A basic list of items that visitors might need is presented in Fig. 4. The host herbarium should advise the visitor on curatorial processes such as gluing slips on specimens or refiling them. Institutional policies should determine the privileges that visitors can be granted concerning accessing the institutional database.

1 Herbarium tour. **2** Artist at work. **3** Researcher Subhani Ranasinghe working with a specimen. **4** Visitors' kit to include: determination slips (archival paper), archival adhesive, scrap pad, archival ink pen, pencil, rubber, sharpener, paper clips, ruler, forceps, mounted needle.

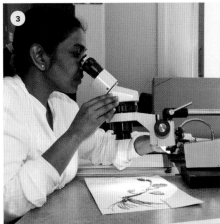

POST-VISIT
Carmen Puglisi and Alan Paton

Once the visit is over there may be outstanding activities necessary for both the visitor and curator to complete. Best practice steps may include incorporating specimens, applying any taxonomic changes suggested by the visitor and supplying feedback on the experience.

Post-visit clean-up work may include re-curation of specimens.

Curation

At the end of the visit, there will likely be curation work to be done. In many cases, this will be a simple matter of reincorporating specimens in the collection. Researchers who introduce changes, such as a new determination or a type designation, should be asked to set aside the specimens for checking before reincorporating specimens in the collection (see: Incorporation). The curator will also update the database accordingly and implement the changes across any associated collection items (carpological, spirit, tissue, seed, living plant). Other requests such as loans or destructive sampling will generally be fulfilled according to the herbarium policies and/or material transfer agreements.

Publication

It is customary for authors to acknowledge the host herbaria in their outputs. When significant resources are deployed in support of the visitor's work, curators should consider negotiating a more prominent role (e.g., co-authorship) in publications resulting from research in their herbarium.

Researchers will likely continue to use the data gathered during their visit long after its end. Curators might want to ask visitors to update them with any taxonomic changes concerning the specimens studied.

Feedback

Feedback could be requested via a questionnaire given to the visitor at the end of their stay or as a periodic exercise where recent visitors are contacted and asked about their experience. It is useful to keep at least a record of the visitor's name and their affiliation, date of the visit, number of days, nature of work and a brief description of collections studied. This information is useful in planning for future visits and providing statistics on the use of the collection, although care must be taken to follow the relevant national laws regarding holding personal data.

General feedback questions on the service provided:

1. Why did you choose to visit this collection?
2. Did the collection meet your requirements?
3. Were the infrastructure and facilities in good working order?
4. Were the staff helpful?
5. Did the use of our collection lead to any outcomes?
6. Were you given enough information to navigate the herbarium building?
7. Was it easy to book a visit?
8. How could the herbarium improve its accessibility and facilities?
9. Would you visit again? Why (not)?

HERBARIUM HIGHLIGHT:
AUSTRALIAN NATIONAL HERBARIUM (CANB)

Brendan Lepschi

Based in Canberra, The Australian National Herbarium is part of the Centre for Australian National Biodiversity Research. The Centre is a joint venture between the Commonwealth Scientific and Industrial Research Organization (CSIRO) and the Director of National Parks, Australian Department of Climate Change, Energy, the Environment and Water (DCCEEW).

Collections and buildings

The Australian National Herbarium (CANB) holds c. 1.2 million specimens, comprised of c. 298,000 non-vascular plants and fungi, and c. 915,000 vascular plants (angiosperms, gymnosperms, pteriodophytes and allies). CANB holds approximately 11,500 types which are digitised and available online; approximately 80% of collections are databased.

Buildings

Founded in 1930, The Australian National Herbarium is located on two adjacent sites on Black Mountain in central Canberra. Angiosperm collections are housed on the CSIRO Black Mountain Science and Innovation Park site, with fungi, non-vascular plants, gymnosperms, pteriodophytes and allies housed on the Australian National Botanic Gardens (ANBG) site. Incorporated herbaria include ANUC, CBG and FRI.

Important Collectors

L.J. Brass, M.I.H. Brooker, R. Brown, N.T. Burbidge, L.A. Craven, J.A. Elix, T.G. Hartley, R.D. Hoogland, M. Lazarides, A.H.S. Lucas, E. Phillips, R. Pullen, R. Schodde, H. Streimann

Scope

Worldwide, with a particular focus on Australia, New Guinea and the South-West Pacific. CANB holds extensive and important collections from New Guinea (c. 213,000 specimens), comprehensive collections of lichens and bryophytes and vouchers for living plant collections held at the ANBG. CANB accessions around 10,000 specimens a year.

Research

The Herbarium is involved in several areas of research, including taxonomy, biodiversity informatics (management and development of the Australian Plant Name Index and Australian Plant Census), systematics, evolution, ecology, conservation and biology of Australian plants. It also provides tools, datasets and expertise to more applied projects on biosecurity, biodiscovery and climate change.

Visitors

Specimens are sent on loan to other herbaria, and around 100 scientists from national and international institutions visit CANB to study the collections each year. Destructive sampling is permitted, at the discretion of the Curator. An annual traineeship programme focuses on curatorial and taxonomic skills and activity.

Environment

Collections are held in five collection halls in three separate buildings on two sites. All halls are environmentally controlled, and fire detection and suppression systems are present in all halls.

Angiosperm collections at CANB housed on the CSIRO Black Mountain Science and Innovation Park site. **1 & 2** CSIRO site New Wing, 1994. **3 & 4** ANBG site, 1974. **5** CSIRO site Old Wing, 1974. Fungi, non-vascular plants, gymnosperms, pteriodophytes and allies housed on the Australian National Botanic Gardens site.

SCIENCE COMMUNICATION
Simon Watt

As the inspirational American poet and civil rights activist Maya Angelou put it, "people will forget what you said, people will forget what you did, but people will never forget how you made them feel". It is here that story-telling becomes important. We should want people we share things with, to come to share at least some of our passions and interest.

A conversation, not a lecture
The specimens that herbaria house tell many tales if only we have the eyes to see them. Examining their history, the context of their collection and their place in our world today all present us with many starting points for stories. The best engagement, though, does not simply seek to transmit these tales, but to be a two-way process, a conversation where both parties learn. If you want to know how to hook your audience, one of the best ways to find out is to ask them. A date palm might mean one thing to the curator of an herbarium but something completely different to a gardener, or to a religious scholar or to a confectioner. Each of their perspectives would enrich the discussion for all. Such a holistic approach is not just enriching in both directions, it is also more educationally effective. Research consistently shows that when people take the voyage of discovery themselves, they are more likely to remember it. If, rather than tell someone the simple facts behind the wind dispersal of seeds, we show them a sycamore key and ask them what they think the wing is for, so they might take that intellectual flight independently, then the germ of knowledge and curiosity is more likely to take root.

Every specimen tells a story
All good stories are tales of conflict. In their simplest form they involve a character that wants something, meeting an obstacle and the solution, or lack of, that results. That character can be anything from a climate-resilient *Coffea* species, rediscovered and developed into a coffee crop species, to a scientist learning from traditional knowledge of medicinal plants for the treatment of malaria, based on information on herbarium specimens. To these basic elements we can add more spice. All stories must have a beginning, a middle and an end, but they do not necessarily have to be in that order. Looking through historic collections, it's easy to see that we can start with where we are now and work backwards to uncover exactly how we got here. To buy more emotional investment from the audience we should detail the stakes involved. What can happen to the food web if one critical thread is snapped? What does it mean for the endangered species when the seeds won't germinate? What exactly is the battle that we are engaged in when looking to plants to find chemicals that could cure cancer?

Sharing is caring
In such a beautiful collection it is easy to get the impression that the herbarium aesthetic is the full picture, but that is very far from the truth. It is an obligation, though a joyous one, to share the collection and what has been learned from it as widely as possible. The herbarium is a site of active enquiry; a hub for study, learning and research that has a deep impact on the wider world. On issues such as food security and the climate crisis, the research going on in herbaria the world over remains strikingly relevant. It is important that

researchers make their voice heard. Herbaria may be like libraries made up of dead material, but interest in them remains very much alive.

KNOW YOUR AUDIENCE

Chelsea Snell

Before you even begin to communicate, understand who your audience is. What is their background? Do they have any understanding of science? This is extremely important in determining how you tell the story.

Sometimes the hook can be hidden. It's important to try hold off the finer details and the facts initially and think about the broader context of the tale you are trying to tell. Put yourself into the headspace of the audience and ask yourself 'why does this matter to them?'

Always start with the key point, or the 'wow factor' of the story then weave in the details. Bring the story to life and aim to invoke the senses by using images or props, where possible.

Personal anecdotes can be powerful in storytelling so strive to humanise and convey your enthusiasm to promote your passion.

Unless your audience is science-informed, avoid using jargon and scientific words.

REACHING YOUR AUDIENCE

Chelsea Snell

There are a variety of ways and platforms in which you can communicate. Define your audience first to allow you to select the right channel.

Digital

Social media channels are free to set up and can be used to reach multiple audiences. For example, Twitter (X) is good for initiating dynamic conversations; Facebook can connect you with local communities, volunteers and citizen scientists; YouTube can allow you to dive into detail. However, it's important to remember that social media can be time-intensive when it comes to building a following, and maintaining engagement and planning is key.

Online blogs offer an outlet for communicating to the scientific community. Many sites offer 1-year free trials and low-cost personal options, e.g., WordPress. Similarly to social media, online blogs need time investment. Online forums such as Reddit (e.g., r/Herbarium and r/botany) present a simpler option.

In person outreach

Connect with local schools, community groups, colleges and universities. Design tours specific to your audience and incorporate activities where possible. Can you build a volunteer group to assist?

FURTHER READING

Pollan (2001); MacGregor (2010–2020); Mars (2010–present); Gottschall (2012); Storr (2019).

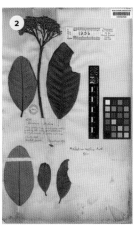

1 *Coffea arabica* cultivated for the production of coffee.
2 *Aspidosperma excelsum* the bark of which is used in traditional medicine for the treatment of malaria.

PUBLIC ENGAGEMENT AND OUTREACH
BEHIND THE SCENES TOURS
Clare Drinkell

Herbaria are unique environments. Not usually public-facing spaces, herbaria can be creatively utilised to offer the wider audience a privileged and memorable learning experience. Behind the scenes tours are an opportunity to improve oral presentations, share ideas, and fascinate visitors about botanical and scientific endeavours.

1 Herbarium tour being given by Sophie Richards. She is holding a picture of *Uvariopsis dicaprio*, a species newly described to science and collected at Ebo forest (a Tropical Important Plant Area in Cameroon). 2 Creating a 'Floras' display for a group tour.

Purpose of a tour

The tour objective is to engage audiences through guided inquiry, specimen-led displays, and conversation. It is helpful to explain how and why a collection is used and showcase examples of current, collections-based research projects. Tours can be an opportunity to gauge public opinion and stimulate curiosity with your audience. It's also useful practice for improving public speaking skills for curation staff. The audience in turn should gain an increased knowledge of the scope, purpose and benefits of the herbarium.

Different types of tours

- Virtual tour
- Members welcome tour
- Community learning programme
- Sensory guided tour
- Specialist interest group
- School group
- University group
- Volunteer tour
- Funders tour

Considerations

Prior to organising a tour, it is worthwhile assessing risks and anticipating hurdles when showing a group around the herbarium. Constricted spaces in the collections may impede access, e.g., wheelchair users. Navigating visitors around working staff may limit numbers, so consider group sizes when publicising tours and account for staff time and resources. Smaller groups are likely to be more engaged and allow for more interaction, e.g., object handling or inviting questions. Facilitate volunteer staff with supervising roles, e.g., health and safety instructions.

Ideas for a display

- Current fieldwork
- New species discoveries
- Uses of geo-referenced herbarium data
- Conservation projects
- Mounting botanical specimens
- Making botanical illustrations
- Integrated Pest Management
- Digitisation projects
- Historical collections
- Collector focus
- Anniversary themes

Tour outcomes

Herbarium tours need to be beneficial for the institution, staff and audience alike. Organising and running tours can be time-consuming; however, educating visitors in the context of the herbarium can be an extremely engaging and immersive learning experience. Advocating and profile-raising of the collection has potential for generating income through ticket sales, fundraising donations, and an incentive to join membership schemes.

Tour guiding is an opportunity to challenge preconceived ideas such as 'plant blindness' (when focus is often on iconic mammals or birds and the importance of plants is frequently ignored), and to use audiences to gauge opinions and improve public engagement.

FURTHER READING
Ashby (2018); Gallimore & Wilkinson (2019).

SCIENCE FESTIVAL
Gemma Bramley

Science festivals offer herbaria the opportunity to engage visitors with the cultural, historical and scientific significance of their collections, associated curatorial activities and their collections-based research.

Science festivals and herbaria
Science festivals offer people opportunities to get involved in a range of experiences led by science professionals to showcase and celebrate all aspects of science. Their often 'hands-on' format gives scope for a style of events that allow more abstract concepts to become concrete and reveal scientific activities that are typically behind closed doors; they are increasing in popularity as their positive role in public engagement with science has become evident (Durant 2013).

Creative learning
Adults, children and families choose to attend science festivals to interact with scientists through programmes specifically designed for them and tend to leave with a heightened interest and curiosity in science (Jensen & Buckley 2014). Herbaria, their collections and the curatorial and research activities occurring within them are ideal subjects for science festivals events. Offering a closer look at herbarium-based activities and preserved collections that are not often public-facing through hands-on workshops, talks and 'behind the scenes' tours raises awareness of herbaria and their role in understanding and conserving biodiversity.

Case study: Kew Science Festival
Adults and children engage with curation and taxonomy through interactive workshops and herbarium tours at Kew's popular Science Festival (see Fig. 1 and 2). There's the opportunity to collect and press plants, and in a 'Pop-up herbarium', visitors may create and digitise their own plant specimen. Participants become 'taxonomists in training' by naming and classifying a selection of plants on display under expert guidance.

1 Identification station: observe and match the leaf to the specimen then create a plant family name.
2 Pop-up herbarium: prepare a specimen, scan and image it.

EDUCATION
Yvette Harvey

The first herbarium c. 500 years ago was specifically created as an educational tool by Luca Ghini. His pressed specimens enabled the teaching of morphological characters to students in the winter when fresh plant material was unavailable (Findlen 2017; Flannery 2018, 2023). To this day, herbaria are still used as such although their educational scope has been much expanded.

The herbarium allows opportunity to reach audiences in a wide range of ways

The 'eureka' moment
Pre-school children through to senior school teenagers will use the collections to explore subjects such as science, art, sociology, geography and history in addition to international concerns such as environmental issues. That elusive but memorable 'eureka' moment can be initiated from a specimen, a note on the label, or even a little bit of research about the specimen.

A teaching tool
For higher-level students studying morphological characters (via the naked eye, a hand lens, microscope or within cells), ethnobotany, classification, conservation (preservation of collections and IUCN red listing), collections management, history, social history, etc.

Work experience
As a coaching environment to discover more about museum curation such as documentation, preventative conservation and adding to the collection through collecting and specimen preparation. It is also used for creative audiences interested in pressing plants or using pressed plants as source material for artwork, poetry, music, literature, etc.

Herbarium display
A free-choice learning environment can be immersive and reactive (Maina 2015). Displays are likely to contain specimens with related items and multi-level messages for the self-guided. If temporary, displays can be of a current world event, anniversary or unusual discovery that will spark an interest and show how useful the collection is to visitors.

Talks from the collections
If space is not a constraint, as a place for public talks about your research, the research of the department, collections stories. Exploring a subject from a different angle may enable viewers to question their perception of a topic.

> **TIP**
> Audiences can engage in learning activities such as pressing, mounting, filing or research.

1 At RHS Garden Wisley, the working and collection areas of the herbarium are all visible to the public. Fixed displays (including opening drawers) have been built into a wall that divides a visitor space from the herbarium suite. The actual herbarium store area's interior windowsills also provide an area for reactive, changing displays while an exterior cabinet is filled with specimens that the public are encouraged to touch. **2** Visitors can watch specimens being created and also have the opportunity to interact with staff. **3** Looking into the public space from RHS Garden Wisley's herbarium. **4** Milwaukee Public Museum has dioramas that capture life in the field. **5** Innovative displays like the ones in the Manx Museum offer a free-choice learning experience for visitors.

PLANT FAMILY SORTS

Sue Frisby

Newly accessioned specimens need to be named or the determination name verified before adding the material to the collections. Identification is easier prior to mounting, when all aspects are visible, dissectible, and examinable with a microscope. A family sort (pre-identification) functions as a valuable teaching and learning opportunity, thus transferring knowledge to the next generation of botanists.

The family sort

Staff, students, or visitors examine a bundle of specimens, with the determination label obscured, around a large sorting table. Participants note characters such as: leaf arrangement, floral parts and fruit shape. Access to a selection of reference books is useful for checking ideas and possibilities (see further reading). Each person has an opportunity to describe their observations, contribute ideas, and suggest the plant family a specimen belongs to. The determination label is then revealed, creating dissension and discussion, or a nod of approval. Queries can be noted in pencil on the label – for subsequent checking.

Techniques for learning family traits

Recognising character patterns in combination with other traits helps lead the thought processes to certain families. For example, opposite leaf arrangement with gland dots and a collecting vein is typical for Myrtaceae. With frequent, regular participation in family sorts, it is possible to get a real feel for the features differentiating the plant families. Online image-based plant resources and field guides are useful tools to aid identifications, as are regional Floras such as: Malesiana, Neotropica, Ecuador, West Tropical Africa, Gabon, Tropical East Africa, Zambesiaca, etc.

TIP

- The process of group family sorts provides an impetus to really examine the material – it is this ability to look that reveals the clues to identification

- The family sort process is a great opportunity for visiting specialists to impart knowledge and methods of familiarising features to help identify plant families

---

FURTHER READING

Guides include: Corner & Watanabe (1969) and Utteridge & Bramley (2015) for tropical plants; Gentry (1996) for the Americas; Hawthorne & Jongkind (2006) for Africa; and van Balgooy *et al.* (2015) for South-East Asia.

1–2 All surfaces are available for close observations. **3** Sort participants examine and discuss characters, before assigning to a plant family. **4** Example page of *Dissotis leonensis* from an online family sort. Online sorts can be shared more widely, enabling botanists from further afield to attend and lead sessions.

THE 'HERBARIUM IN A BOX' IDEA

Nina Davies and Olivia Porritt

Herbaria are not widely known outside of the botanical world. Access to most physical collections is restricted, with access only for research purposes or open to a wider group on the odd special occasion such as scheduled tours or open days. The 'Herbarium in a Box' idea is a way of taking specimens and associated objects out of an herbarium to engage with the public and increase awareness of these collections and the science connected to them. There are many uses of an herbarium (Funk 2003b); each specimen is unique and can be used to explain plant science, information about collectors or about a specific time in history.

Creating a 'Herbarium in a Box'

Events such as science festivals have proven a popular means of bridging the gap between science communication and public engagement (Bultitude *et al.* 2011). Using herbarium objects to engage audiences can help to inspire careers in botanical science and increase awareness of the importance of these collections.

The 'Herbarium in a Box' can be any size as long as it's portable and will also depend on the objects which are chosen for engagement. For example, at Kew, objects included herbarium material, therefore a green herbarium specimen box was chosen to ensure these fit within the box.

Ideas of objects to include in a 'Herbarium in a Box':

Think about the audience you want to engage with or the event, for example, school groups, local communities or at a science festival. This will help to guide which herbarium objects to include.

Ideal objects include those which help to explain collections-based botanical science, stories from particular botanists or species. Think about objects which are relatable.

Object examples are listed below:

- Photographs – find photographs of collecting trips for example, botanists in action, living plants, habitats or landscapes which help to explain fieldwork

- Fieldwork equipment – helps to explain how botanists collect herbarium material; object examples include a collecting book, leech socks, a small press and a hand lens

- Herbarium specimens and bulky fruits – as handling is a risk to specimens, it is best to create new specimens for this purpose. Choosing local plants which tell an interesting story or something recognisable, for example, when dried, *Theobroma cacao* fruits smell like chocolate

- Be aware of permissions when collecting and do not use anything restricted, poisonous or which cannot be handled easily

- Laminated specimens – laminated pressed and dried plants are robust, waterproof and smudge-proof

1 The 'Herbarium in a Box' used to engage groups at Kew Gardens, showing a range of different objects which all fit into an herbarium green box. **2** Laminated plant specimens – making them robust for handling. **3** Choosing botanical objects that can be handled.

Case study: using the 'Herbarium in a Box' to engage with groups at RBG Kew

Professional development training workshops for youth group leaders

Outline

Professional training days were organised to encourage youth group leaders and youth-focused charities to visit Kew Gardens with their service users and introduce them to learning resources. A youth group leader is someone who facilitates and delivers workshops and sessions to young people outside of mainstream educational settings, often to marginalised groups of young people who face some form of disadvantage.

Using the 'Herbarium in a Box'

The 'Herbarium in a Box' was used as part of an interactive session to help youth group leaders develop their understanding of Kew Gardens as a scientific organisation as well as a beautiful garden.

Objects from the box were shared with the group to prompt a discussion on where the objects came from and the story of the plant. For example:

- The Madagascan periwinkle – discussion about medicinal plants
- Cacao pods – discussion about food and economic importance of plants
- The Kew team also used the objects to explore what a herbarium is and why Kew's Herbarium is important around the world

Outcomes and impacts

Feedback from many group leaders found that:

- They had not realised that Kew Gardens was a scientific organisation, and getting to see real specimens from the herbarium helped them connect to the science in a more personal way
- That this interactive session made them feel more confident to visit Kew Gardens with their young people and talk more about plants, conservation and science

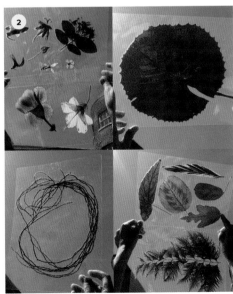

HERBARIUM HIGHLIGHT: CAMBRIDGE UNIVERSITY HERBARIUM (CGE)

Lauren M. Gardiner

The Cambridge University Herbarium (CGE), founded in the 1760s, holds 1.1 million plant and cryptogams with an estimated 50,000 types. Collections are global, with particularly rich 19th-century-type material and 20th-century British specimens (Walters 1981).

Background

Representing over 350 years of plant collection, exchange, and research by alumni, staff, and an extensive network of correspondents of the University, by the late 20th century, CGE had become poorly known, relatively inaccessible, and was little used in modern research or teaching activities. New leadership has focused on strategic alignment of activities with current University research, teaching, and public engagement; building links with other collections, museums, and libraries across the University; and bringing curation and collections care in line with international standards and best practice (Gardiner 2018).

Important collectors

Darwin, Henslow, Lindley, Babington, Hooker, Martyn, Rackham, Spruce, Wallace, Wallich.

Research and teaching

Making sure the Herbarium contributes to and supports University-based research, teaching, and public engagement has been essential to secure institutional support and long-term funding. Pilot projects with engaged researchers, undergraduate and postgraduate students; getting several lectures, tours, and practical classes into the undergraduate curriculum; and small-scale digitisation work utilising small grant funds all feed into the development of a strategic and fundraising plan for CGE, and help to raise the collection's profile inside and outside the organisation.

Public engagement and outreach

Initial digitisation and imaging of specimens is focusing on known type specimens and specific historically important collections (such as the Darwin and 18th-century Martyn collections; see Rose & Gardiner 2021). These specimens are the most commonly requested by researchers and are relatively easy to build engaging stories around to communicate the importance and modern use of collections to the public. A university setting presents many different ways to engage the public. Such communication is a vital component of modern research funding. Herbarium specimens provide visual, tangible, and relatable materials that can connect the viewer to concepts and research activities. Even with very limited resourcing and staffing, and no public display space of its own, CGE has been able to offer in-person and virtual online tours of the collection, private views, and provide specimens, artworks, and archive materials for display in temporary and permanent exhibitions and events at the publicly accessible museums and libraries in Cambridge. Public lectures and herbarium tours at annual local events (e.g., Cambridge Festival, Open Cambridge, Alumni Festival), and displays and demonstrations at the Festival of Plants and Dissertation Fair reach different audiences, as have a YouTube channel, Twitter account, and the development of a new CGE website. Local radio, television, and newspaper features have raised interest in the collections, and initial forays into opportunities for 'digital volunteers' to transcribe specimen labels via citizen science platforms have proved popular and will be expanded in the future.

1 Bust of John Stevens Henslow whose 19th-century collections are included in CGE Herbarium. **2** Open compactor shelving of the herbarium. **3** Digitisation of select specimens is intended to increase accessibility and raise the profile of the collections. **4** The holotype and only record of the species *Sicyos villosus*. This specimen was collected by Charles Darwin on expedition to Galápagos, 1835, and is now thought to be extinct. **5** A university setting helps to engage with the public and communicate the modern use of collections. **6** The herbarium contributes to and supports University-based research.

HERBARIUM SNAPSHOTS
HERBIER NATIONAL DE GUINÉE
Dr. Sékou Magassouba

Created in 2009, the National Herbarium of Guinea (HNG) resulted from the memorandum of understanding of collaboration between Gamal Abdel Nasser University, Conakry (UGANC) and the Board of Directors of RBG Kew, through the dedication of the researcher professors and Researchers of the Chair of Eco-botany (Faculty of Sciences).

Research priorities
A public institution of a scientific and educational nature, the National Herbarium of Guinea (HNG) has a mission to contribute to the implementation of national government policy through research and training (Masters) in the field of botany.

Our research projects focus on threatened and endemic plants and identifying Tropical Important Plant Areas (TIPAs) (Couch *et al.* 2019) and how best to protect them. We partner with many organisations on our projects, including Royal Botanic Gardens, Kew, local NGOs and national governments.

Botanical treasures
Guinea has the greatest plant diversity in West Africa with many rare and endemic species. However, the rapid development of mining projects in recent years increasingly threatens these species through degradation and loss of their habitat. There is an urgent need to establish effective scientific means or methods to prioritise new conservation efforts to safeguard this unique and endangered plant diversity.

Public engagement is factored into our projects as most Guineans have no idea of the national treasure of interesting and endemic species, e.g., *Pitcairnia feliciana* is the only member of the Bromeliaceae in Africa.

National flower campaign
In 2018, we organised the campaign to choose a national flower from among 16 threatened species. A series of workshops in the capital towns of the four natural regions were held with well-defined terms of reference. Public high schools and communities were the target audiences for these workshops, and a social media campaign targeted a larger audience. At the end of the vote, the voters returned *Vernonia djalonensis*, the "Chardon du Fouta" as "National Flower" of our country (Couch 2018).

We have also run citizen science workshops, teaching people how to collect an herbarium specimen and providing information posters on species that are important for conservation, so if they find them, they can let us know.

> **TIP**
> Pictures and photographs work really well to get the message across, as well as having enthusiastic presenters. It is also important to know who your target audience is, so that you can adapt your presentation accordingly and present it in the local language.

1 Dr Martin Cheek teaching botanical survey techniques to HNG Masters students. **2** *Pitcairnia feliciana*, the only Bromeliaceae in Africa. **3** Workshop with teachers at Kindia for the National Flower Campaign. **4** Voting for the National Flower at the Student Forum 2018. **5** Threatened species poster for *Ternstroemia guineenis*.

⁵ Avez-vous vu cette plante ?
Aidez à sauvegarder votre patrimoine guinéen!

Ternstroemia guineensis Cheek

Statut globale (UICN) : En Danger

Cette espèce se trouve uniquement en Guinée. Nouvelle espece en 2019

Arbuste à 5 m de haut ou rarement un arbre à 9 m des forêts submontagnardes; écorce du tronc et des gros tiges épais, gris, en mosaïque. Fleurs hermaphrodites, 7 – 9 mm diamètre, pendantes. Sépales blanches, pétales jaunes ou jaune-blanchâtres. Fruit sur un pédicelle rouge, pendantes, 2 – 5 par rameaux feuillus sous-tendu par les sépales persistantes rose-verdâtres.

Saison de floraison : novembre à février
Saison de fructification : mai

Menaces : Les incendies sont faits pour le pâturage. La fréquence de ces incendies modifie la qualité de l'habitat à la lisière des galeries forestières sous-montagnardes.

Si vous la trouvez, contactez nous!
Dr Sékou Magassouba à l'Herbier National de Guinée, UGANC, T: 622 278185, nationalherbierguine@yahoo.fr
Mr Mamadou Diawara à Guinee Ecologie T: 621 27 75 08, info@guineeecologie.net
Mamadou Saliou Kanté à AGEDD T: 622 18 05 21, mskantef2013@gmail.com

Pour nous aider à confirmer son identification, prière de bien vouloir: 1) prendre une portion de feuilles, tige et fleurs; 2) faire sécher et placer entre deux feuilles de papier blanc; 3) noter la date, le lieu de récolte, votre nom et votre contact; 4) Contacter les personnes ci-dessus pour toute question.

Franklinia   Kew

MAKINO BOTANICAL GARDEN HERBARIUM
Dr. Kazumi Fujikawa

The Kochi Prefectural Makino Botanical Garden (MBK), which honours the remarkable achievements of prominent botanist Dr. Tomitaro Makino from Kochi, was opened to the public in 1958. With over 3000 plant species on display, the eight hectares of gardens offer colour and interest throughout the year. The herbarium was established in 2000; currently, it houses around 320,000 specimens of vascular plants. Research is focused on the flora of Kochi and Myanmar, and around 10,000 specimens are added to the collection every year.

Contribution towards the flora of Myanmar
In 2000, MBK launched a joint research project with the Forest Department, Myanmar, combining a floristic inventory and subsequent economic botanical development for the conservation of natural resources in Myanmar. At present, over 32,500 herbarium specimens from Myanmar have been collected, which are used in taxonomic research by professional botanists for a better understanding of Myanmar flora. The joint project provides training sessions for government staff who are engaged in capacity-building in plant inventory and conservation, and supports local community development programmes to achieve some of their Sustainable Development Goals.

Kochi flora and our volunteers
The MBK acts as Kochi's regional education centre for taxonomic research and plant diversity. MBK houses more than 120,000 specimens from the whole Kochi region, and they are collected by our devoted volunteers. The MBK regularly holds workshops for the volunteers to improve their understanding of the collection as well as species identification. More than 200 volunteers have been working with us to develop our research on the flora of Kochi.

Our outreach programmes
'Travelling Makino' is one of our public outreach programmes. Staff travel to various towns to give lectures on plant science in an informal setting over a cup of coffee. They also conduct specific outreach activities including events for visitors to the garden through exhibitions; for example, an exhibition titled 'Herbarium' included a display of featured specimens and research with a demonstration of how to mount specimens. 'Collection Rescue Operation' is the third example of our outreach programmes. Since 2019, MBK has been taking part in this operation as a member of the nationwide network of herbaria in Japan. We are rescuing damaged specimens from heavy floods in Kyushu district by cleaning and restoring them in our herbarium.

1 The Herbarium. **2** Inventory in Myanmar. **3** Workshops in the field for volunteers. **4** Special exhibition about the herbarium. **5** Demonstration of how to mount specimens. **6** Cleaning and restoring specimens.

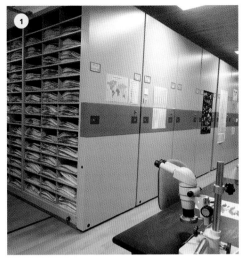

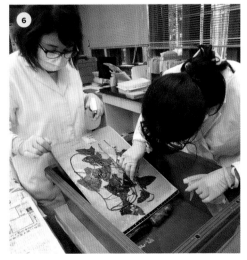

HERBARIUM OF THE CALIFORNIA ACADEMY OF SCIENCES
Sarah Jacobs, Emily Magnaghi and Nathalie Nagalingum

The California Academy of Sciences in San Francisco's Golden Gate Park houses a combined public museum space, aquarium, and research collections. Founded in 1853, our herbarium (CAS) has the largest collection of vascular plants in the western U.S. with over 2.3 million specimens.

Tours and public talks
Public engagement and outreach are critical to our work; they draw interest in the collections, our research, and to plants themselves – all of which maintains the relevance and importance of the collection as well as promoting plants and fighting 'plant blindness'. At CAS, we engage largely through tours, talks, and museum events where we highlight details of the specimens, curation, and research, emphasising the wealth of information that herbaria contain.

Tours provide a direct glimpse into the workings of an herbarium, and talks showcase the science that collections inform. Both are amenable to pre-recorded, virtual formats.

Favorite audience topics include:

- Notable botanists/scientists
- Economically and culturally significant plants
- Unique/unusual plants
- Expeditions and forays (historical and contemporary)
- Taxonomic collections
- Ongoing research

Display and information tables
Our museum hosts themed events (e.g., seasonal or special interests such as women in science), and we participate by creating a table that highlights plants and our collections correspondingly. Our tables aim to include appealing components for a variety of age groups, utilising multiple communication tools and techniques to share information that is inclusive and accessible to a broad audience.

Example components:

- High-resolution, scanned replicas of our specimens (to avoid damage from handling)
- Corresponding living plant displays from local gardeners/plant shops
- Phone adapters for dissecting scopes, or digital microscopes connected to laptops to increase ease of access to small details on plants/specimens
- Project images and videos on larger monitors/TVs for easy viewing
- Handouts/colouring sheets about plant biology
- Items for "busy hands", e.g., make-your-own stickers or stamps

Future efforts
Opportunities for digital engagement with the public continue to expand, providing new avenues for outreach. We are actively working to incorporate digital tools such as Notes From Nature to engage the public in label transcription. We also use digital tools such as iNaturalist to document plants when appropriate (e.g., BioBlitzes), generating interest in plants and the work we do. Opportunities such as these elevate plant awareness and broaden the reach of the herbarium into diverse communities.

1–3 A carnivorous plant display. **1** Digital scans from the herbarium and corresponding living specimens. **2** Living plants available for touching and holding, and a dissecting scope with close-up viewing. **3** A TV monitor displays associated video compilations and table for make-your-own stickers. **4–6** Examples of content displayed during public tours of the collection. **4** The tools of a field botanist. **5** Plant collections that become specimens. **6** Specimens, images, and maps illustrating the breadth of a given botanist's travels.

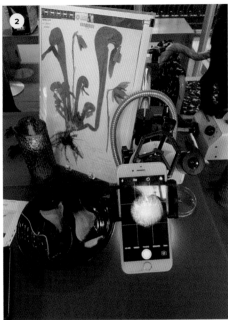

ROYAL BOTANIC GARDEN EDINBURGH
Robyn Drinkwater

The Royal Botanic Garden Edinburgh's Herbarium (RBGE) contains nearly 3 million specimens, representing from a half to two-thirds of the world's flora. The collection covers over 300 years of biodiversity, with the oldest specimen collected in 1697. The herbarium supports a wide range of activities to increase public engagement with the collections.

Frankenstein's Plants
Frankenstein's Plants was run as part of the Edinburgh Science Festival, to increase the awareness of the Herbarium at Edinburgh and the research it helps to support.

Participants were able to make their very own unique herbarium specimen using a range of pressed and dried flowers, leaves and fruits that had been bought in or collected from the garden at Edinburgh. They also "digitised" their specimens to see and share the results in an on-line gallery.

Activity stations
A series of activity stations allowed participants to create specimens using a range of techniques similar to those employed in an herbarium.

Through these stations 'collectors':

- Selected material for the specimen
- Mounted it onto an herbarium sheet using gummed tape
- Completed a label with collection information including the collector's name and collector number. A plant description could be added using basic descriptive terms for flowers and leaves
- Name given to their 'species' using a list of real and imagined genera and species epithets
- Imaged the specimen, which was later added to a virtual herbarium – a Flickr site (called Frankenstein's Plants) set up for the event

The completed specimens could then be taken home by the participants.

Aims and outcomes

Our hope with this event was to engage visitors in the work that we do. The aims of this event were to:

- Make the public aware that the RBGE has an herbarium and explain what it is and how it is used by scientists
- Show how specimens are created by attaching pressed and dried plants to card
- Show that botanical names come in two parts: genus, species, and an introduction to other botanical terms
- Raise awareness of the digitisation of the collections
- As it was an all-ages event, but run within a children's science festival, it allowed active engagement with the children and tailored outreach to accompanying adults

Over the course of three days, 333 specimens were created by 'collectors' ranging in age from 3 to 95. Since the event, there have been over 11,000 views of the images in the Flickr Frankenstein's plants virtual herbarium. The event was very successful, and we plan to run it again in the future.

1 Stations for creating a specimen. Clockwise: pressed material selection, naming and description, imaging, information about the herbarium. **2** 'Collector' arranging and mounting their specimen. **3** Arranging and mounting specimens. **4** 'Collectors' and their completed specimens. **5** Finished specimen *Spotianum explorii*. **6** Finished specimen *Ceryskiana griffindori*.

245

SINGAPORE BOTANIC GARDENS HERBARIUM

David J. Middleton, Serena Lee and Bazilah Ibrahim

The herbarium of the Singapore Botanic Gardens (SING) was founded in 1875, and today houses around 750,000 specimens, including about 10,000 type specimens, primarily from Singapore, Peninsular Malaysia, and other parts of South-East Asia. It is a centre of research on the plant diversity and ecology of Singapore and the wider region.

UNESCO World Heritage Site

Singapore Botanic Gardens is the world's only tropical botanic garden with UNESCO World Heritage Site status, inscribed in 2015. The historical and current research programmes of the Gardens, along with the Gardens' landscapes and place in the social fabric of Singapore, helped to ensure the bid was successful. It is equally important that the Gardens also engage with local policy makers, academics and the general public to ensure this legacy is understood and protected.

Tours and identifications

Once per month on a Saturday, the Gardens offers a tour of the Herbarium and other research facilities for the public. Research and curatorial staff take it in turns to conduct these tours to explain the purpose and work of the Herbarium, concentrating on how the collections enable us to better understand the plant and fungal diversity of Singapore.

The National Parks Board (NParks) is the parent body of Singapore Botanic Gardens and has a remit to conserve biodiversity, manage the nature reserves and enhance green spaces in the country. All staff have a tour of the Herbarium as part of their induction programme, as do students for some courses in the local universities and polytechnics.

Knowledge of the herbarium is also enhanced through active volunteer and intern programmes.

The Herbarium plays an important advisory role in identifying both wild and cultivated plant species from Singapore. This is a free service for the general public, but there is a charge for those who require identifications for professional purposes.

Media

The Gardens' status as Singapore's first World Heritage Site and its position within NParks makes it a favourite with broadcast, social and print media. The herbarium and its staff have consequently featured in many television programmes and newspapers, in both the Herbarium itself and in the field while collecting. Similar to many major herbaria, SING also often hosts visits for VIPs.

The Gardens publishes a magazine, *Gardenwise*, that features articles from across the range of activities of the Gardens and always has features on the work of the herbarium written in an accessible way for the general public.

1 A tour for the general public, called the Race Against Time tour, conducted once per month on a Saturday. **2** A photographer with plant mounter Suganthara Davi for a media event. **3** *Gardenwise*, the magazine of Singapore Botanic Gardens. **4** Former Secretary-General of the United Nations, Ban Ki-moon, visiting the Herbarium and Library with former Director of Singapore Botanic Gardens, Dr Nigel Taylor.

HERBARIUM HIGHLIGHT:
WILLIAM AND LYNDA STEERE HERBARIUM (NY)

Barbara M. Thiers

The New York Botanical Garden, including The Steere Herbarium (NY), was founded in 1891. The Herbarium holds approximately 8 million specimens of Algae, Bryophytes, Fungi (including Lichens) and Vascular Plants with approximately 150,000 type specimens.

1 William and Lynda Steere Herbarium, New York Botanical Garden. **2** Herbarium Imaging Center, New York Botanical Garden.

Buildings

The Steere Herbarium is housed in a collections facility that it shares with the Garden's LuEsther T. Mertz Library. The Herbarium occupies the lower four floors of the building, and the library collections are on the top floor. Designed for 20 years of growth, the building opened for use in 2002. In addition to the specimen ranges, the building has suites of visitor offices on three floors. Each range contains additional workspace on half-height cabinets, and internet access via wifi is available throughout the Herbarium. The arrangement of specimens is currently being updated to reflect the relationships suggested in APG III.

Important collections

Foundational collections include the herbaria of John Torrey (vascular plants), Job B. Ellis (fungi), August Jaeger and William Mitten (bryophytes), and Frank S. Collins (algae). Staff collections of note include those of Nathaniel and Elizabeth Britton, William Murrill, Per A. Rydberg, Henry Gleason, Henry. H. Rusby, G.T. Prance, Arthur Cronquist, and Patricia and Noel Holmgren, to name a few. The Steere Herbarium has incorporated a number of the herbaria; among the largest of these came from Columbia, Princeton, and De Pauw Universities.

Scope, (etc.)

The focus of the Steere Herbarium has traditionally been on the plants and fungi of the Americas, with special emphasis on specimens from Eastern and Intermountain West portions of North America, the Caribbean, and Brazil. In recent years, a significant number of specimens from Australia (fungi), South-East Asia and the Pacific Region (vascular plants) has been added as the result of staff research projects.

Research

The Steere Herbarium supports the research of Garden scientists who conduct biodiversity, genomic and ethnobotanical research throughout the Americas, Asia and the Pacific. The Herbarium also supports research by scientists worldwide; its specimens are cited in an average of 250 publications per year. Digitised sections of the Herbarium include all type specimens and algae; most bryophytes and fungi; and about 60% of vascular plants. All incoming accessions are digitised. All digitised data are shared online through the Garden's C. V. Starr Virtual Herbarium (http://sweetgum.nybg.org/science/) and through a variety of other data portals.

Visits and loans

The Steere Herbarium welcomes an average of 230 researchers per year for in-person visits. Such visitors spend on average 2000 days per year consulting the collection. Additionally, we have about 200,000 online visitors to our C. V. Starr Virtual Herbarium per month. We send on average 25,000 specimens on loan per year. Instructions for prospective borrowers or visitors can be found at the Steere Herbarium website. Where applicable, Material Transfer Agreements and other provisions of the Nagoya Protocol govern the research use of specimens.

Environment

All accessioned specimens are held within the climate-controlled collections building, with the exception of some recently acquired herbaria. The herbarium of the Brooklyn Botanical Garden (BKL), which is on long-term loan to the Garden, occupies a wing outside the collections building. Collections in process are stored in a cold storage room on the first floor of an adjacent building.

H.423/85
20
KEW

Colchicum szovitsii Fischer & C. A. Mey

det./conf. K. Persson   3/5 1

Colchicum hydrophilum Siehe

FLORA OF TURKEY

C6 Hatay: Amanus Mts. Karlik Tepe, from
Sogukoluk. Hillsides: limestone:seasonally
waterlogged grassy meadows. Few scattered.
Alt: 1100m
Flowers: pink.

26 March 1967

M.J.Cheese   C.M. & W. 2491

COLCHICUM SZOVITSII
Fisch et. May.

DET C.D.Brickell 2.2. 19 82

GLOSSARY

Sue Frisby and Nicola Biggs

This Glossary attempts to define terms and abbreviations found throughout the handbook, together with some definitions of more common words and terms used more generally in an herbarium. The Glossary is quite specific for herbarium terms, and additional sources should be referred to for other phrases and words used in botany and taxonomy; for example, morphological terms used in plant taxonomy can be found in Beentje (2016); a list of the most common symbols, units and abbreviations used in plant-biological research can be found in Planta 213: 158–162 (Anonymous 2013); shape definitions are found in the Systematics Association Committee (1962); plant names and their etymology etc. are covered in Stearn (2004).

ABS Clearing House	Access and Benefit-Sharing Clearing-House, a website administered by the CBD Secretariat. https://www.cbd.int/abs/absch
accession	The legal entry of a specimen into an institution. The process by which the herbarium commits to the long-term care of a specimen (see Spectrum). Legal title of the specimen will have been granted to the institution prior to accessioning (see also acquisition).
acid-free paper	Papers made with alkaline pulp, so the pH is neutral, including compounds which neutralise acids from the atmosphere and natural aging processes, recommended for herbarium paper and determinavit slips.
acquisitions	'According to Spectrum 5.1 standards' means that legal title of the specimen has been transferred to the holding herbarium. However, the term may be insufficient to convey the ownership and status of the specimens under access to generic resources legislation, where legal ownership of genetic resources may by retained by the country of origin. In these instances, an herbarium may commit to the long-term care of the specimen, i.e., incorporate it, without owning it, and therefore neither the term acquisition nor accession is appropriate for long-term acceptance of such material. Another term such as custodianship could be used in these cases.
adsorb/adsorption	The adhesion of atoms or molecule, i.e., water to a surface, creating a film on the surface of the absorbent, e.g., silica gel.
aggregator	A website or program that collects related items of content and displays or links them, e.g., Google News.

ALA DigiVol	Atlas of Living Australia, DigiVol is a virtual citizen science project which allows people worldwide to participate in unlocking biodiversity data from a range of data resources; data includes herbarium collection labels, field notebooks and camera trap images. https://www.ala.org.au/
algorithm	A process or set of rules to be followed in calculations, especially by a computer.
allergens	An antigen causing a strong immune response.
altimeter	An instrument for determining altitude (elevation above sea level).
ambient	Surrounding completely, encompassing; used to refer to the immediate surroundings of something in the herbarium.
ammonia	A compound of nitrogen and hydrogen, it is a colourless gas or liquid with a pungent smell, used as a softening agent for dried plant material.
ancillary collections	Any related collections other than an herbarium sheet, such as: material in spirit, seed collections, bulky/carpological collections, timber collections, illustrations and economic botany collections.
anoxic	The absence of oxygen. Via a specialised chamber, useful for killing pests on delicate herbarium specimens.
APG	Angiosperm Phylogeny Group. https://www.mobot.org/MOBOT/research/APweb/
appr. (Latin)	Approbavit – he or she confirmed; means the same as confirmavit (qv.).
apud	With, among or in. Formerly used in author citations for 'in' (in the writings of).
archival	Paper, glue and pens of a quality that is permanent and resistant to any deterioration.
Arctos	A data capture and management system for museum collections. http://arctosdb.org
argon gas	An inert gas used as a fire suppressant by displacing atmospheric oxygen.
asphyxiated	Deprived of air. A risk to personnel associated with the use of a clean agent (e.g., argon) as a fire suppressant.
autoclaving	Treatment with high temperature and pressure of 121 degrees C for 15 minutes; used for sterilisation.
barcode	A machine readable code, used as an individual specimen identifier.
bench fees	Payment made for use of equipment or facilities.
BICON	Biosecurity Import Conditions, an Australian Government Department database. https://bicon.agriculture.gov.au
bilateral	Relating to two sides, or involving two parties.
binocular microscope	Optical microscope with two eyepieces allowing use of both eyes simultaneously.
biocides	Substances or chemicals which destroy harmful organisms.
biocultural	Ethnobiological or biocultural artefacts documenting human interactions and uses of plants and animals.
biodiscovery	Research on samples or extracts of biological resources for genetic or chemical purposes for actual or potential value to humanity.

biodiversity	The variability among living organisms from all sources including, inter alia, terrestrial, marine and other aquatic ecosystems and the ecological complexes of which they are part; this includes diversity within species, between species and of ecosystems.
bioprospecting	Looking for commercially valuable products from plant or animal species.
biosecurity	Procedures or measures designed to protect the plant/animal/human population against harmful biological or biochemical substances.
blotting paper	Highly absorbent paper.
boiling ring	Electrical appliance for boiling water in herbaria, used to rehydrate dried plant material.
BRAHMS	Botanical Research And Herbarium Management System: database software for managing taxonomy. https://herbaria.plants.ox.ac.uk/bol/
bryophytes	Mosses and liverworts.
buckram	Stiff cotton cloth, with a loose weave.
bulky	Larger three-dimensional collections of plants, cross-referenced, and kept as a separate collection when too large for the herbarium sheet.
bund	A wall around storage, to protect from unintended spillage or escape.
bundles	Herbarium bundles are usually 50 specimens tied up between cardboards with string, secured for easy, safe movement.
camphor	A volatile, crystalline substance from the wood of *Cinnamomum camphora* used as an insect repellant.
capsule	A folded archival paper envelope, glued onto the herbarium sheet, used to store small pieces of plant material for subsequent examination or destructive sampling, including for DNA extraction.
carpological collection	The seeds and fruits of plants, cross-referenced, and kept as a separate collection when too large for the herbarium sheet.
casings	Shed skin casings of growing insect larvae.
CBD	Convention on Biological Diversity. https://www.cbd.int/
centipedes	Predatory arthropods, with one pair of legs per body segment, and a venomous bite.
CETAF	Consortium of European Taxonomic Facilities. https://cetaf.org/
charophytes	Freshwater green algae.
checksum	A small-sized block of data derived from another block of digital data for the purpose of detecting errors that may have been introduced during its transmission or storage.
CITES	Convention on International Trade in Endangered Species of Wild Fauna and Flora. An international agreement between governments aiming to ensure that trade in specimens of wild animals and plants does not threaten the survival of the species. https://cites.org/eng
clinometer	An instrument for measuring the angle or elevation of geographical slopes.
cloth tape	Gummed linen tape.
cloud storage	Online data storage.

coll. — Collegit = he or she gathered, placed before the collector's name on a data label. Also used as an abbreviation for 'collector'.

combination — A second author has revised the taxonomy of a taxon and moved it into a new genus, creating a new combination.

comm. — Communicavit = he or she communicated. Used on data labels to indicate the person who sent the specimen to the herbarium; this is usually not the person who collected it.

compactor — A mobile storage system, with shelving mounted on mobile bases which slide along rails, optimising available storage space.

compartmentation — Dividing something, such as a building, into separate sections.

conf. — Confirmavit = he or she confirmed; printed or written on the determination slip, followed by the name of the botanist who is agreeing that the specimen is correctly named.

consignment — A batch of items, usually referring to bundles of herbarium specimens being sent to or received from other herbaria.

CoP — Conference of the Parties, e.g., CBD CoP, where countries come together as a group for decision making. https://unfccc.int/

Copenhagen mix — Solution for storing preserved plant material made up of industrial methylated spirit, distilled water and glycerol in a ratio of 70:28:2.

Correx — Corrugated plastic: an extruded, twin-wall, fluted polymer that can have additives to make it UV resistant, flame retardant or biodegradable.

corrugates — Sheets of e.g., aluminum, pressed on a corrugating machine, placed between cardboard sheets and specimens in flimsies in the press, enabling airflow from a dryer to pass through the press, speeding the drying process.

COSHH — Control of Substances Hazardous to Health (U.K. Govt regulations). https://www.hse.gov.uk/coshh/

data loggers — An electronic device for recording data, usually left in place to record temperature over a long period.

dataset — A collection of separate sets of information that is treated as a single unit.

D.B.H. — Diameter at Breast Height. This is a standard method of measuring the diameter of the trunk or bole of a standing tree.

deaccession — The formal decision by a governing body, or on its behalf, to take objects out of its accessioned collection.

derivatives — Something based on another source; plant-based derivatives refer to ingredients or materials originating from plants such as bioactive compounds (vitamins, minerals, tannin, flavonoids), colourants, fibers, etc.

Derma Shield — Branded hand cream, lasts up to 4 hours on the skin and through washes, and protects against irritants, e.g., chlorine, industrial dirt.

determinavit slips — Paper labels used when annotating herbarium sheets, with the identification and identifier's name and date; often abbreviated to 'det. slips'.

detritus — Decaying organic matter.

diagnostic — A distinctive characteristic useful for identification.

digitisation	The conversion of text, specimens or images into a digital form that can be processed by a computer.
diorama	A replica or model of a scene.
disposal	The official mode of transferral. Most commonly via gift, exchange, destruction.
DNA	Deoxyribonucleic acid – genetic instructions in the form of a double helix.
DNA sequences	Nucleotide order in DNA.
DoeDat	Digitale Ontsluiting Erfgoedcollecties, funded by the Flemish Government under a project called DOE! – Digital Access to Cultural Heritage Collections. DoeDat is all about creating data and "doe dat" means "do that" in Dutch. https://www.doedat.be/
downstream digital analyses	Transmission of data to an end user.
drying oven/field stove	A transportable means of generating heat with which to dry plant specimens in the field.
drying paper	Paper to absorb moisture from drying plant material.
DSLR	Digital Single-Lens Reflex camera.
duplicate	Several sheets of a single collection, illustrating all available characters, distributed to other herbaria.
EarthCape	A biodiversity database platform to capture, organise and analyse data. https://earthcape.com
economic botany collection	A collection of plant materials and artefacts of economic importance to man.
ecosystem services	Benefits to humans from environmental ecosystems.
elytra	The hardened fore wings of beetles and earwigs.
ethanol	Alcohol, an organic chemical compound used in herbaria to preserve plant material, often used in collecting when floral structures might be difficult to interpret once pressed flat.
ethnobotany	The study of interrelations between humans and plants.
Europeana	A web portal created by the European Union containing digitised cultural heritage collections from institutions across Europe. https://www.europeana.eu
ex num. or e	e numero = from the number. The specimen has been named by tracing the collector's name and number in published lists. This particular duplicate has not been annotated and probably not seen by the author of the revision. There is therefore always a slight possibility that owing to an error, e.g., in numbering, the specimen is not correctly named.
ex situ	Off-site, processes or things away from the natural location.
exchange	A consignment donated with an expectation of something in return, i.e., specimens or identifications.
exsiccata	Refers to a set of herbarium specimens usually provided with printed labels; mostly accurately named sets of duplicates chosen to be uniform and representative of the taxon (e.g., 'Schedae ad Herbarium florae U.S.S.R.').

exsiccatae	A list of specimens seen, which is often published in revisions or monographs.
exsiccatus	Dried plant and fungal herbarium specimens.
FADGI	The Federal Agencies Guideline for Digitisation. A collaborative effort by USA agencies to create a set of technical guidelines, methods, and practices for digitising historical, archival and cultural content. http://www.digitizationguidelines.gov
Fairtrade	A certification system ensuring high-quality, ethically produced products that are traded at fair prices.
fecundity	The ability to reproduce.
field press	Lightweight plant press used to flatten and dry specimens in the field.
fixative solution	A solution that preserves cells and tissue structure, e.g., formaldehyde.
flimsies	Lightweight sheets of paper to absorb moisture from specimens in the press; newspapers (tabloid size) are best; also papers used to protect individual herbarium sheets when going on loan.
forceps	Fine straight or curved biology tweezers.
formaldehyde	A colourless pungent gas.
formalin	A colourless solution of formaldehyde in water.
forsan	Means perhaps. Used to indicate a doubtful determination.
frass	Pelleted larval excreta; may appear powdery.
freeze-dried	Preserved by rapid freezing followed by treatment under a high vacuum.
fume cupboard	A well-ventilated enclosure to extract volatile chemicals.
fungal hyphae	The long filamentous branches found in fungi.
GBIF	Global Biodiversity Information. An international network and data infrastructure aimed at providing open access to data about all types of life. https://www.gbif.org/
genomic	The study of genes and chromosomes.
georeferencing	To provide latitude and longitude points for localities given on specimen labels.
germplasm	Living tissue containing the genetic material from which new plants can be grown.
gift	A consignment of herbarium specimens freely given from one herbarium to another.
GIS	Geographic Information System, used for creating, mapping, managing and analysing geographic information. https://www.esri.com/en-us/what-is-gis/overview
glass slide	Glass microscope slide.
glassine	Thin, fairly strong, glazed and translucent paper used to make envelopes for storage, e.g., delicate plant parts or as a "window" over unglued plant parts.
glycerol	A simple polyol compound, colourless and non-toxic, it has antimicrobial properties, and is used to stop material drying out.

gnaw	To persistently bite or chew and eventually eat away.
GPS	Global Positioning System, a satellite-based navigation system used to plot exact geographic locations.
grapnel	A hooked anchor or clamp, used for grasping or holding something.
graticule	A grid of intersecting latitude and longitude lines, which can be housed in an eyepiece, calibrated and used for measuring.
hand-lens	A magnifying glass designed to be held in the hand, usually x10 magnification.
haud	= not at all; used to disagree, usually with a determination.
humidify	Increase the level of moisture in the air.
hydrogen phosphide	A poisonous gas used as a fumigant.
hygrometer	An instrument for measuring humidity.
ICN	International Code of Nomenclature, which sets rules and recommendations that govern the scientific naming of all organisms traditionally treated as algae, fungi or plants. https://www.iapt-taxon.org/nomen/main.php
identification	The determination of a plant or taxon (such as species or subspecies) as being identical with or similar to another and already known element. In some instances, the plant may be found to be new to science.
iDigBio	Integrated Digitized Biocollections. Data and images are being curated, connected and made available in electronic format. https://www.idigbio.org
illustration collection	A collection of illustrations housed within the herbarium sequence, used for plant identification.
in	'in' frequently connects two personal names, e.g., *Viburnum ternatum* Rehder in Sargent. Here the first author validly published the name but in a work otherwise written (or edited) by a different author. Although citations using 'in' are often written as author citations, they are in fact incorrect and it is better to write '*Viburnum ternatum* Rehder in Trees & Shrubs [Sargent] 2: 37 (1907).'
iNaturalist	A social network for naturalists to record their observations. https://www.inaturalist.org
indigenous	Native.
ingress	The action of entering.
invasive species	An introduced species causing harm in its new habitat.
Iron Gall ink	A purple-black ink made from iron salts and plant-based tannic acids.
IUBS-TDWG	International Union for Biological Sciences–International Working Group on Taxonomic Databases for Plant Sciences. A standard database format for recording data on the use of plants. https://iups.org
IUCN	International Union for the Conservation of Nature. Composed of both government and civil society organisations, it is the authority on the status of the natural world and the measures needed to safeguard it. https://www.iucn.org

JABOT Rio de Janeiro Botanic Garden Herbarium Collection, an institutional dataset system for plant collection and management. https://servicos.jbrj.gov.br/v2/jabot/

Japanese tissue A strong, thin tissue made from vegetable fibres.

jeweller's tags Small white, strung ticket ideal as a number tag for specimens in the field. See also number tags.

Jiffy sheets Sheets of polyurethane packing foam.

JPEG A format for compressing image files. A commonly used method for compressing images, which can be adjusted to allow a trade-off between storage size and image quality.

Jstor Journal Storage. Provides access to millions of journal articles, books, images and primary sources including Global Plants with thousands of online images of herbarium specimens worldwide. https://www.jstor.org

Kew mix Solution to rapidly kill plant tissue to preserve structure, made up of industrial methylated spirit, formalin, glycerol and distilled water in a ratio of 53:5:5:37.

Kraft Union paper Branded, waterproof paper for protecting items in transit.

lacquer A hard protective coating, using sap from a tree or synthetic substances as a varnish.

larvae Juvenile form of an insect (also nymph).

larval skins A moulted exoskeleton.

LED Light-Emitting Diode, a semiconductor light source that produces light when a voltage is passed through it. LEDs have advantages over incandescent light sources, such as lower power consumption, longer life and smaller size.

leech socks Large, fine-textured, tight-knit fabric socks to wear over your boots, to prevent leeches attaching themselves.

legislation This is the process or product of promoting national laws by legislature, parliament or another governing body. It may have many purposes: to regulate, to authorise, to outlaw, to provide funds, to sanction, to grant, to declare or to restrict.

Les Herbanautes A citizen collaborative digital herbarium developed by the Paris Herbarium from the photos of the plants of the Paris Herbarium and the network of French naturalist collections. http://lesherbonautes.mnhn.fr/

Libsorb A softening and wetting agent, used to soften dried plant material for dissection.

Loan In an herbarium setting this refers to botanical specimens being dispatched to another institution following a request. An institution will have a set of procedures and paperwork to follow.

Lossless File Types Allows the original data to be perfectly reconstructed from the compressed data with no loss of information, e.g., RAW and TIFF files (NB. TIFF files can also be lossy).

loupe	A small hand-held magnifying glass, especially those designed to fit the eye socket, for seeing small details more closely. They generally have a higher magnification than a magnifying glass. On some loupes the lens folds into an outer housing that protects the glass.
Lycophytes	One of the oldest lineages of living vascular plants, e.g., clubmosses, selaginellas.
MAA	Material Acquisition Agreement; contains the terms and conditions under which materials are acquired by an institution and their use once acquired.
Machine Learning	A form of Artificial Intelligence that makes predictions from data by using algorithms and statistical models. It allows software applications to become more accurate at predicting outcomes without being explicitly programmed to do so.
mainframe	A computer used for bulk data processing; it has more processing power than servers, workstations and personal computers.
mercuric chloride	A heavy, highly toxic, white, odourless crystalline compound of mercury and chloride used as a disinfectant and fungicide.
Metamorfoze	Netherlands national programme for the preservation of paper heritage. The programme concentrates on material of Dutch origin that is kept in cultural heritage institutions such as libraries, museums and research institutes. https://www.metamorfoze.nl/english
methylated spirits	Ethyl alcohol denatured with methanol. A solvent comprised of 90% ethanol and 10% methanol, sometimes with other additives, mainly to make it unsafe for drinking by introducing an unpleasant taste, smell or to cause nausea.
MGNSW	Museums & Galleries of New South Wales, Australia. Helps, by means of grants or expertise, small-medium museums, galleries and Aboriginal cultural centres create exciting experiences for visitors and, through this, thriving NSW communities. https://mgnsw.org.au/
microfiber (or microfibre)	A very fine synthetic fibre having a diameter of less than ten micrometres (one millionth of a metre), most commonly made from polyester or nylon. The shape, size, and combinations of synthetic fibres are chosen for specific characteristics, e.g., softness, toughness, absorption, water repellence, electrostatics, and filtering ability.
misidentification	A name that has been applied to the wrong taxon or accidently to the wrong sheet/specimen by human error when adding determination slips.
morphological	Relating to external structure or form, specifically in biology the form of living organisms and the relationships between their structures, in order to determine their function, development and how they may have been shaped by evolution.
morphometric	The process of measuring the external shape and dimensions of an object and the relationship between size and shape.
mounted needle	Very fine dissecting needles that are mounted into a thicker handle, either metal or wood, for ease of use. Can be used to hold a specimen in place or to move parts around.

mounted specimen	Plant specimens attached to mounting sheet/board either with archival glue, paper strips or stitches of cotton/linen thread, to protect the specimen from damage. Ideally the mounting sheet/board would be acid-free. Well-prepared and cared-for mounted specimens will last many years.
mounting sheet/board	Strong paper or board/card, robust enough to support a specimen attached to it; ideally it should be of archival quality and acid-free.
mounting specimens	The act of permanently gluing with archival adhesive (or other method, e.g., paper strips), a dried, pressed plant specimen onto a sheet of ideally archival paper or card.
MSA	Material Supply Agreement; the terms under which materials are supplied/provided (whether gift or loan) to another institution or individuals.
MTA	Material Transfer Agreements; contains the terms under which materials are transferred (whether gift or loan) to an institution.
Nagoya Protocol	An International legal framework that enables equitable sharing of genetic material, including traditional knowledge and the benefits that arise from their use. The name was adopted from the 2010 meeting in Nagoya, Japan. https://www.cbd.int/abs/about/
Nalgene	Brand of plastic products designed originally for laboratory use that are lightweight and shatterproof.
naphthalene	A white crystalline, water-insoluble hydrocarbon used as a repellant, distilled from coal tar, and used in mothballs, for example.
nematodes	Round worms and eelworms (plant-parasitic nematodes), found in most environments: marine to fresh water, polar regions to the tropics.
Nitrile	Synthethic rubber material used to make gloves which offer resistance to chemicals and abrasion.
non	= not
notho-	Is used to indicate a taxon (of any rank) known to be of hybrid origin, e.g., nothospecies, nothovariety. etc.
noxious substances	Any substance or material that is potentially harmful to life.
number tags	A strung, white card tag used to write collection numbers on specimens before being pressed. See also jeweller's tags.
nymph	The juvenile form of an insect (also larva) which undergoes gradual change before reaching the adult form.
OCR	Optical Character Recognition. The identification of printed characters using photoelectric devices and computer software. Four top features of OCR are accuracy, ability to convert images of text into editable text, ability to convert text from different languages and a fast processing speed.
palynology	The study of plant pollen, spores and microscopic plankton.
parasitoids	Insects whose larvae feed and develop within or on the body of another species, eventually killing their host organism.

parthenogenic	Pertaining to parthenogenesis; asexual reproduction whereby offspring are produced from an egg alone without sperm.
pathogens	Microorganisms which cause disease, such as viruses, bacteria or fungus.
pesticides	Chemicals or substances used to control or kill pests; these include insecticides, herbicides and fungicides. Most are used on crops or cultivated plants.
petri dish	Shallow dish made of glass or plastic, often with a lid, used mainly to hold a growth medium in which cells may be cultured.
phenological	The study of periodic events and how they are influenced by seasonal changes, variations in climate or habitat factors, e.g., the date of flowering or emergence of leaves.
pheromones	Chemical substances released by an insect or mammal, affecting the physiology or behaviour of another organism.
phylogenetic system	A system based on the evolutionary history and patterns of relationships between organisms. For example, plant families arranged so that closely related families are stored together.
phytosanitary	Relating to plant health, particularly with respect to the legal requirements of international trade and accompanied by a government inspection certificate to show that a shipment has been treated or is free from pests and diseases before being imported or exported.
pipette	A slender tube, often of glass, used to measure small amounts of liquid.
plant blindness	An informally proposed form of cognitive bias by humans towards animals, and a tendency to ignore plant species, such as not noticing or recognising the importance of plant life in an environment.
Plastazote	A branded, specialist form of closed cell foam for packaging. Structurally it is made up of a series of enclosed air pockets (cells) that do not interconnect, so that the foam is harder, firmer and tougher than open-celled foam. Benefits include resistance to liquid, shock absorption and thermal insulation.
plaster beetles	Beetles in the family Latridiidae, measuring about 1 to 2 mm in length, that thrive in damp plasterwork.
plasticiser	A substance (usually a solvent) added to a material to make it softer and more flexible, to increase its plasticity or to decrease its viscosity. Usually added to plastics and rubber.
Pohl's solution	A softening and wetting agent, used for making dried flowers etc. pliable for dissections; named after Dr. Richard Pohl. Wetting solutions contain active chemical compounds that reduce surface tension to induce optimal spreading of a fluid.
Polish Presses	Plant presses made of welded steel mesh.
polyester	A synthetic resin used to make synthetic textile fibres, and ultimately fabrics/sheeting that can be used for packaging/bags to store specimens in an herbarium or in the field.
polyphagous	Able to feed off many different types of food.
polyethylene	A tough, light, flexible synthetic resin used for packaging.

PPI — Pixels Per Inch, a measure of resolution in image files; the greater the number of pixels the greater the detail in the image.

preservative — A substance used to preserve against decay.

press straps — Straps (usually 2) which hold a plant press together under pressure, to flatten the contents. Woven straps with spiked buckles provide the tightest grip.

pressing — Preserving plant material by placing between sheets of paper and applying weight and/or pressure, to flatten and dry the specimen.

protocol — The official procedure or system of rules governing a formal agreement, treaty or record of an agreed method between parties.

pupa — Non-feeding stage in the life-cycle of insects undergoing complete metamorphosis (for moths = chrysalis).

pyrethroid — A synthetic, organic compound related to natural pyrethrins, with insecticidal properties for household and commercial use.

QR code — A machine readable matrix barcode, composed of black and white squares. Often containing data for a locator, identifier, or tracker that points to a website or application.

RAW file format — Files containing uncompressed and unprocessed image data; this format stores the largest amount of detail that can be edited, compressed or converted and saved to other formats.

Reflora — Repatriating Brazil's plant diversity information and building capacity for greater use through digitisation, dissemination and research visits. http://reflora.jbrj.gov.br

relative humidity — The amount of water vapour present in air expressed as a percentage of the amount needed for saturation at the same temperature.

respirator — An apparatus worn over the mouth, nose or the entire face to prevent the inhalation of dust, smoke, or other noxious substances.

RNA — Ribonucleic acid, present in all living cells, functions as a messenger from DNA and controls the synthesis of proteins.

rodenticide — A chemical poison made and used for killing rodents.

salvage — To save items, that are often valuable, from damage or destruction or to conserve items that have been damaged, for example by fire or water ingress.

SBSTTA — Subsidiary Body on Scientific, Technical and Technological Advice. Article 25 of the CBD establishes an open-ended intergovernmental scientific advisory body known as SBSTT to provide the COP and, as appropriate, its other subsidiary bodies, with timely advice relating to the implementation of the Convention. https://www.cbd.int/sbstta/

scalpel — A small, extremely sharp-bladed metal instrument with replaceable blades, used for dissecting plant parts or for fine cutting of paper/card.

sched. — Scheda = label. Used in literature to indicate information taken from an herbarium label (in sched. = on an herbarium label). See also 'exsiccata' above.

Schweinfurth method	A method of preserving botanical specimens in the humid tropics. Recommended by Schweinfurth, tied bundles of collected specimens between cardboards, are placed upright in a bag or other container, and alcohol is poured over the specimens until everything is thoroughly moistened. The bundle(s) are then wrapped and packed as firmly as possible for transport. The specimens remain soft, pliable and moist for years and can be dried at the collector's convenience. https://archive.org/details/jstor-2477267/page/n1/mode/2up
secateurs	A pair of pruning clippers, for use in one hand.
section symbol §	Silcrow §, section of a genus.
seed bank	A store for seeds so as to preserve genetic diversity for the future, a form of insurance against loss of plant species.
seismic	Relating to earthquakes and vibrations in the earth's crust.
Sellotape	Transparent adhesive tape sold in a roll and used for joining things together, such as paper or card.
sens	(sensu) = in the sense of. If in literature this term is placed between a specific epithet and an author's name it implies that that author used the name incorrectly, perhaps as the result of a misidentification. Such names are usually listed with synonyms but they are result of a misidentification, and as a result are not true synonyms. The specific name as used by the original author will belong to a different taxon.
SiBBr	Sistema de Informaçâo sobre a Biodiversidade Brasileira or Brazilian Biodiversity Information System. Integrates data and information about biodiversity and ecosystems from different sources and makes them accessible. https://sibbr.gov.br/?lang=en_GB
Silica gel	An adsorbent form of silicon dioxide, its high specific surface area allows it to adsorb (see also adsorbtion) water readily, making it useful as a drying agent.
SISBIO	Sistema de Autorização e Informação em Biodivesidade. SISBIO is a remote assistance system that allows researchers to request authorisations to collect biological material and to carry out research in federal conservation units and caves. https://www.gov.br/icmbio/pt-br/servicos/sistemas/sisbio-sistema-de-autorizacao-e-informacao-em-biodiversidade
SISGEN	Sistema Nacional de Gestão do Patrimônio Genético or National System of Management of Genetic Heritage and Associated Traditional Knowledge. An on-line system to register genetic heritage, whether access to, samples of, final products or accreditation. https://portal.ufvjm.edu.br/prppg/comite-e-comissoes/patrimonio-genetico-sisgen
skeletal leaves	Fleshy part of leaves eaten away, leaving just the venation network behind.
slingshot	A catapult used for collecting tree samples without the need to climb. The projectile is attached to a wire or rope that can be fired into the canopy to detach a plant specimen.
Specify	Specify Collections Consortium processes data associated with specimens in biological research collections. https://www.specifysoftware.org/

Spectrum A collections management standard developed by the Collections Trust (2022; Spectrum 5.1), giving a useful overview of the general collection management processes describing which important information should be recorded. Available online: https://collectionstrust.org.uk/spectrum

spirit Liquid preservative mixtures to store plant material; preserve plant parts in original state, albeit without pigmentation, for study. See also Copenhagen and Kew mix.

spore print Pattern produced when spores from a fungal fruiting body fall onto paper or glass, showing gill position, and colour and distribution of spores.

square brackets [ ] Often used in lists of synonyms to indicate an erroneous use of a name, rather than a true synonym or an illegitimate name (e.g., homonym).

stakeholder A person or organisation affected by or with an interest in the activities at stake. Stakeholders involved in conservation and the granting of collecting permits and prior informed consent for access may include relevant departments of central government, local authorities, private individuals such as landowners, indigenous peoples, local communities, and non-governmental organisations.

substrate The substance or material on which an organism lives and gains nourishment, such as soil, rock, plants, man-made materials.

subvariety A taxonomic rank below that of variety but above that of forma.

Symbiota Open-Source Biodiversity Data Management Software. It is specifically designed towards efficient, collaborative digitisation and an open data exploration and publishing tool, e.g., species inventories, interactive identification keys, taxonomic descriptions, data entry from label images and data publishing to GBIF and iDigBio. https://symbiota.org/

syn. nov. Synonymum novum = new synonym. Used to indicate that the author believes it to be the first time that the name has been treated as a synonym of the present taxon.

TIFF Tag Image File Format. A flexible, adaptable file format for holding multiple images and data within a single file through the inclusion of 'tags' in the file header.

tarpaulin A heavy-duty, waterproof canvas used for protecting exposed areas, objects or people from the elements. Can be attached to standing objects by means of rope to make a temporary roof in the field.

Taxa Plural of taxon.

TDWG Taxonomic Databases Working Group (historically) or Biodiversity Information Standard (today). A not-for-profit, scientific and educational association formed to establish international collaboration between creators, managers and users of biodiversity information and to promote sharing of knowledge. It develops, promotes and ratifies standards and guidelines and acts as a forum for meetings and publications. https://www.tdwg.org/

teste = according to or by the witness of. Sometimes used instead of 'det.'

tethered Imaging Images are saved directly onto a computer's hard drive via a camera, tablet, smartphone etc. through a cable or wireless app.

thermometers	Devices for measuring the temperature of a solid, liquid or a gas; the common units for measurement are Celsius, Fahrenheit or Kelvin.
timber collection	Collecting a sample of tree wood for specimen vouchers, including the bark and noting any characteristics, e.g., smell, exudate, colour banding between older and newer rings. Also, useful for wood anatomy and thin section research. Can also refer to a collection of wood specimens (sometimes called xylarium or xylotheque).
trait data	Recording physical or specific characteristics of an individual; traits can be determined by genes, environmental factors or a combination of both.
transaction	An exchange or interaction between people/institutions: sharing specimens, DNA, published works, expertise, people.
Tyvek	A synthetic non-woven material made of high-density polyethylene fibres, bonded together by heat and pressure. A trademarked product of DuPont, it is widely used for making protective clothing, waterproofing and protective wrapping.
unmounted specimens	Plant specimens not glued onto paper. Preferable for purposes of examination under a microscope and for removing flowers/fruits but as they are dry and brittle they are susceptible to damage. See also mounted material.
URI	Uniform Resource Identifier. A string of characters used to identify the name and location of a file or resource on a computer network, the best known of which is the URL.
URL	Uniform Resource Locator. The address of a web page, a reference to a web resource that specifies its location on a computer network and a mechanism for retrieving it.
vegetation type-physiognomic	Using the general or characteristic form of the vegetation, e.g., grassland, woodland, cerrado etc.
vegetation type-species based	Relying on the occurrence of certain common species to define the vegetation type.
vel	= 'or', often placed before sp. aff. to mean: or related species.
viroids	Infectious pathogens of small, single strands of RNA.
vix	= scarcely. Used, e.g., on determinavit slips and usually initialled. It is placed against a previous determination to indicate that the second botanist disagrees with it.
watch glass	Shallow dish made of glass or plastic, used as a surface to evaporate liquid, to hold solids being weighed or for heating a small amount of material. Glass ones can be reused, while plastic ones tend to be single-use to avoid cross-contamination.
watermark	A faint design made in paper during its manufacture that is visible when held to a light and typically identifies the maker.
webbings	Larval spun silk, found across food sources, often combined with frass.
wetting agent	A solution used to soften delicate plant tissue prior to dissection. See also Pohl's solution and Libsorb.
woodborers	Beetles whose larval or adult forms eat into and destroy wood.
woodlice	A terrestrial isopod crustacean.

EX NASIONALE } HERBARIUM, PRETORIA, Suid-Afrika.
NATIONAL } South Africa.

Prov. *Bechuanaland* Dist.

Ornithogalum seineri (Engl. + Kr.)
Obern. in. (Bothalia 1964).
syn. *Urginea langii* Bsem.
(*Bulbine seineri* Engl. + Kr.)
syn. *Ornithogalum wilmaniae* Leight.
*from Derde poort just over the
Transvaal border. Flowered
at Prinshof, Pretoria,* 30-10-'59

Pre. No. _____ Legit *L. E. Codd*

Alt. _____ Anno. _____ No. *8875*

REFERENCES

A

Alexiades, M.N. (1996). *Selected Guidelines for Ethnobotanical Research: a field manual.* New York Botanical Garden, Bronx, NY.

Anonymous. (2013). Units, symbols, abbreviations. *Planta* 213: 158–162.

Arts Council England. (2014). *Collections development policy template.* Arts Council England. https://www.artscouncil.org.uk/sites/default/files/download-file/Collections_Development_Policy_template_2014_PDF_0.pdf

Ashby, J. (2018). Museums as experimental test-beds: Lessons from a university museum. *J. Nat. Sci. Collections* 5: 4–12.

Ashley, C.W. (1944). *The Ashley Book of Knots.* Doubleday, New York.

B

Baker, W.J. & Dransfield, J. (2006). *Field Guide to the Palms of New Guinea.* Royal Botanic Gardens, Kew.

Baker, W.J., , Bailey, P., Barber, V., Barker, A., Bellot, S, Bishop, D., Botigué, L.R., Brewer, G., Carruthers, T., Clarkson, J.J., Cook, J., Cowan, R.S., Dodsworth, S., Epitawalage, N., Françoso, E., Gallego, B., Johnson, M.G., Kim, J.T., Leempoel, K., Maurin, O., Mcginnie, C., Pokorny, L., Roy, S., Stone, M., Toledo, E., Wickett, N.J., Zuntini, A.R., Eiserhardt, W.L., Kersey, P.J., Leitch, I.J. & Forest, F. (2022). A comprehensive phylogenomic platform for exploring the Angiosperm Tree of Life. *Syst. Biol.* 71: 301–319.

Bakker, F.T., Antonelli, A., Clarke, J.A., Cook, J.A., Edwards, S.V., Ericson, P.G.P., Faurby, S., Ferrand, N., Gelang, M., Gillespie, R.G., Irestedt, M., Lundin, K., Larsson, E., Matos-Maraví, P., Müller, J., von Proschwitz, T., Roderick, G.K., Schliep, A., Wahlberg, N., Wiedenhoeft, J. & Källersjö, M. (2020). The Global Museum: natural history collections and the future of evolutionary science and public education. Peer*J 8:* e8225.

Balick, M.J. & Herrera, K. (2014). Chapter 4: Curating ethnobiological products. In: J. Salick, K. Konchar & M. Nesbitt (eds), *Curating Biocultural Collections: a handbook*, pp.55–65. Royal Botanic Gardens, Kew.

Banks, S.B. (2015). Managing risks from hazardous substances in the Economic Botany Collection at the Royal Botanic Gardens, Kew: a pragmatic approach. *J. Institute Conservation* 38: 130–145.

Bedford, D.J. (1999). Vascular plants. In: D. Carter & A. Walker (eds), *Care and Conservation of Natural History Collections*, pp. 61–80. Butterworth Heinemann, Oxford.

Beeckman, D.S. & Rüdelsheim, P. (2020). Biosafety and biosecurity in containment: A regulatory overview. *Frontiers Bioengineering Biotech.* 8: art. 650.

Beentje, H. (2016). *The Kew Plant Glossary: an illustrated dictionary of plant terms*. Second Edition. Royal Botanic Gardens, Kew.

Bentley, A.C. (2007). *Shipping and Handling of Natural History Wet Specimens Stored in Fluids as "Dangerous Goods" – Hazardous Materials.* National Park Service (NPS) Conserv-O-Gram No. 11/13.

Boone, M.E. & Basille, M. (2019). Using iNaturalist to contribute your nature observations to Science. *EDIS* 2019 (4): 5.

Botanic Garden Conservation International. (2021). *ABS Learning Package [Online].* Available at: https://www.bgci.org/resources/bgci-tools-and-resources/abs-learning-package/

Brown, G., Jobson, P., Milne, J. & Schönberger, I. (2018). ALERT Lessons from a biosecurity disaster. *Biodiversity Information Science and Standards* 2: e25941.

Bultitude, K., McDonald, D. & Custead, S. (2011). The rise and rise of Science Festivals: an international review of organised events to celebrate science. *International Journal of Science Education, Part B: Communication and Public Engagement* 1(2), 165–188.

Bye, R.B. (1984). Voucher specimens in ethnobiological studies and publications. *J. Ethnobiology* 6: 1–8.

C

Canteiro, C., Barcelos, L., Filardi, F., Forzza, R., Green, L., Lanna, J., Leitman, P., Milliken, W., Pires Morim, M., Patmore, K., Phillips, S., Walker, B., Weech, M.-H. & Nic Lughadha, E. (2019). Enhancement of conservation knowledge through increased access to botanical information. *Conservation Biology* 33: 523–533.

Carine, M.A., Cesar, E.A., Ellis, L., Hunnex, J., Paul, A.M., Prakash, R., Rumsey, F.J., Wajer, J., Wilbraham, J. & Yesilyurt, J.C. (2018). Examining the spectra of herbarium uses and users. *Botany Letters* 165: 328–336.

Carter, D.C. & Walker, A.K. (1999). *Care & Conservation of Natural History Collections.* Butterworth-Heinemann, Oxford.

Chatrou, L.W., Turner, I. M., Klitgaard, B.B., Maas, P.J.M. & Utteridge, T.M.A. (2018). A linear sequence to facilitate curation of herbarium specimens of Annonaceae. *Kew Bulletin* 73: art. 39.

Chase, M.W. &Hills, H.H. (1991). Silica gel: An ideal material for field preservation of leaf samples for DNA studies. *Taxon* 40: 215–220.

Cheek, M., Tchiengué, B. & van der Burgt, X. (2021). Taxonomic revision of the threatened African genus P*seudohydrosme* Engl. (Araceae), with *P. ebo,* a new, critically endangered species from Ebo, Cameroon. P*eerJ. 9*: e10689.

Cheek, M., Nic Lughadha, E., Kirk, P., Lindon, H., Carretero, J., Looney, B., Douglas, B., Haelewaters, D., Gaya, E., Llewellyn, T., Ainsworth, A.M., Gaorov, Y., Hyde, K., Crous, P., Hughes, M., Walker, B.E., Campostrini Forzza, R., Wong, K.M. & Niskanen, T. (2020). New scientific discoveries: Plants and fungi. *Plants, People, Planet* 2: 371–388.

Child, R.E. (1994). The Thermo Lignum process for insect pest control. *Paper Conservation News* 72 [9].

Child, R.E. (2011). *The wider use and interpretation of insect monitoring traps.* In: Winsor, P., Pinniger, D., Bacon, L., Child, B., Harris, K., Lauder, D., Phippard, J. & Xavier-Rowe, A. (eds), *Integrated Pest Management for Collections. Proceedings of 2011: A Pest Odyssey, 10 Years Later.* English Heritage, Swindon.

Christenhusz, M.J.M., Reveal, J.L., Farjon, A., Gardner, M.F., Mill, R.R. & Chase, M.W. (2011). A new classification and linear sequence of extant gymnosperms. *Phytotaxa* 19: 55–70.

CITES. (2021). The CITES website. Geneva, Switzerland. Compiled by the CITES Secretariat. [Online]. Available at: https://www.cites.org/. [Accessed 03/09/2021].

Clark, S.H. (1986). Preservation of herbarium specimens: an archive conservator's approach. *Taxon* 35: 675–682.

Collections Trust (2022). *Spectrum 5.1.* https://collectionstrust.org.uk/spectrum/

Consortium of European Taxonomic Facilities (CETAF). (2019). *Consortium of European Taxonomic Facilities (CETAF) Code of Conduct and Best Practice for Access and Benefit-Sharing.* Available online: https://cetaf.org/wp-content/uploads/CETAF-Best-Practice-Annex-to-Commission-Decision-C2019-3380-final.pdf

Cook, D., Lee, S.T., Gardner, D.R., Molyneux, R.J., Johnson, R.L. & Taylor, C.M. (2021). Use of herbarium voucher specimens to investigate phytochemical composition in poisonous plant research. *J. Agricultural and Food Chemistry* 69: 4037–4047.

Corner, E.J.H. & Watanabe, K. (1969). *Illustrated Guide to Tropical Plants.* Hirokawa, Tokyo.

Couch, C.A. (2018). *Guinea: The Campaign for a National Flower.* https://www.kew.org/read-and-watch/guinea-the-campaign-national-flower.

Couch, C.A., Cheek, M., Haba, M., Molmou, D., Williams, J., Magassouba, S., Doumbouya, S. & Diallo, M.Y. (2019). *Threatened Habitats and Tropical Important Plant Areas of Guinea, West Africa.* Solopress, UK.

Creative Commons (2020). About the licenses. https://creativecommons.org/licenses/ [accessed 31/01/2022].

Croat, T.B. (1978). Survey of herbarium problems. *Taxon* 27: 203–218.

Cronquist, A. (1988). *The Evolution and Classification of Flowering Plants,* 2nd ed. NYBG, New York.

D

Dalla Torre, K.W. von & Harms, H. (1900–1907). *Genera siphonogamarum ad systema Englerianum conscripta.* Lipsiae, G. Engelmann.

Darbyshire, I., Manzitto-Tripp, E.A. & Chase, F.M. (2021). A taxonomic revision of Acanthaceae tribe Barlerieae in Angola and Namibia. Part 2. *Kew Bulletin* 76: 127–190.

Davis, C.C., Kennedy, J.A. & Grassa, C.J. (2021). Back to the future: a refined single user photostation formassively scaling herbarium digitization. *Taxon* 70: 635–643. https://doi.org/10.1002/tax.12459.

Desmond, R. (1995). *The History of the Royal Botanic Gardens, Kew.* Harvill Press, London.

Diazgranados, M. & Funk, V.A. (2013). Utility of QR codes in biological collections. *Phytokeys* 25: 21–34.

Digital Preservation Coalition (DPC) (2015). Digital Preservation Handbook, 2nd Edition, https://www.dpconline.org/handbook, Digital Preservation Coalition © 2015.

Dipper, F. (2016). *The Marine World: a natural history of ocean life.* Wild Nature Press, Plymouth.

Djarwaningsih, T., Sunarti, S. & Kramadibrata, K. (2002). *Panduan Pengolahan dan Pengelolaan Material Herbarium serta Pengendalian Hama Terpadu di Herbarium Bogoriense.* Herbarium Bogoriense, Bidang Botani, Pusat Penelitian Biologi – LIPI, Bogor.

Dransfield, J. (1986). A guide to collecting palms. *Annals of the Missouri Botanical Garden* 73: 166–176.

Drinkwater, R.E., Cubey, R.W.N. & Haston, E.M. (2014) The use of Optical Character Recognition (OCR) in the digitisation of herbarium specimen labels. *PhytoKeys* 38: 15–30.

Durant, J. (2013). The role of science festivals. *Proceedings of the National Academy of Sciences.* 110 (8): 2681.

E

Enghoff, H. (2019). *Does digitisation of natural history collections reduce the need for physical access and physical loans?* Report on subtask 3.2.1 under Task 3.2 "Facilitating Access beyond SYNTHESYS3". SYNTHESYS.

ENSCONET. (2009). *ENSCONET Seed Collecting Manual for wild species.* Royal Botanic Gardens, Kew (UK) & Universidad Politécnica de Madrid (Spain). Available at: http://ensconet.maich.gr/PDF/Collecting_protocol_English.pdf

Ertter, B. (1999). Elements of Herbarium Layout and Design. In: D.A. Metsger & S.C. Byers (eds), *Managing the Modern Herbarium*, pp. 119–145. Society for the Preservation of Natural History Collections, Royal Ontario Museum.

F

Findlen, P. (2017). The death of a naturalist: Knowledge and community in late Renaissance Italy. In: G. Manning & C. Klestinec (eds), *Professors, Physicians and Practices in the History of Medicine*, pp. 127–167. Springer, New York.

Flannery, M.C. (2018). *Blog: At the Beginning: Luca Ghini.* https://herbariumworld.wordpress.com/2018/03/05/at-the-beginning-luca-ghini/

Flannery, M.C. (2023). *In the Herbarium. The Hidden World of Collecting and Preserving Plants*. Yale University Press, New Haven CT.

Forman, L.L. & Bridson, D. (1989). *The Herbarium Handbook*. Royal Botanic Gardens, Kew.

Forrest, L.L., Hart, M.L., Hughes, M., Wilson, H.P., Chung, K.-F., Tseng, Y.-H. & Kidner, C.A. (2019). The limits of Hyb-Seq for herbarium specimens: impact of preservation techniques. *Front. Ecol. Evol.* 7: 439.

Frick, H. & Greeff, M. (2021). *Handbook on natural history collections management – A collaborative Swiss perspective*. Swiss Academies Communications 16 (2).

Funk, V.A. (2003a). The Importance of herbaria. *Plant Sci. Bull.* 49 (3): 94–95.

Funk, V.A. (2003b). 100 Uses for an herbarium (well at least 72). *A.S.P.T. Newslett.* 17 (2): 17–19. Available online at http://www.virtualherbarium.org/vh/100UsesASPT.html

Funk, V.A. (2018). Collections-based science in the 21st century. *J. Syst. Evol.* 56: 175–193.

G

Gallimore, E. & Wilkinson, C. (2019). Understanding the 'Effects of Behind-the-Scenes' tours on visitor understanding of collections and research. *Curator: The Museum Journal* 62: 105–115.

Gardiner, L.M. (2018). Cambridge University Herbarium: rediscovering a botanical treasure trove. *J. Natural Science Collections* 6: 31–47.

Gentry, A.H. (1996). *A Field Guide to the Families and Genera of Woody Plants of Northwest South America.* The University of Chicago Press, Chicago.

Germán, M.T. (1986). Estructura y organización del herbario. In: A. Lot & F. Chiang (eds), *Manual de herbario*, pp. 11–30. Consejo Nacional de la Flora de México.

Gold, K. (2014). *Post-harvest handling of seed collections.* Technical Information Sheet 4. Available at: http://brahmsonline.kew.org/Content/Projects/msbp/resources/Training/04-Post-harvest-handling.pdf

Gottschall, J. (2012). *The Storytelling Animal.* Houghton Mifflin Harcourt Boston.

Gómez-Bellver, C., Ibáñez, N., López-Pujol, J., Nualart, N. & Susanna, A. (2019). How photographs can be a complement of herbarium vouchers: A proposal of standardization. *Taxon* 68(6): 1321–1326.

Grenda-Kurmanow, G. (2021). Adhesives used in herbaria: Current practice with regard to what we know from written sources on mounting herbarium specimens and conservation. *Taxon* 70(1): 1–15.

Guiry, M.D. & Guiry, G.M. (2017). *AlgaeBase*. World-wide electronic publication, National University of Ireland, Galway.

Güntsch, A., Hyam, R., Hagedorn, G., Chagnoux, S., Röpert, D., Casino, A., Droege, G., Glöckler, F., Gödderz, k., Groom, Q., Hoffmann, J., Holleman, A., Kempa, M., Koivula, H., Marhold, K., Nicolson, N., Smith, V.S. & Triebel, D. (2017). Actionable, long-term stable and semantic web compatible identifiers for access to biological collection objects. *Database (Oxford)* Volume 2017, 2017, bax003.

H

Hardy, H., van Walsum, M., Livermore, L. & Walton, S. (2020). Research and development in robotics with potential to automate handling of biological collections. *Research Ideas and Outcomes* 6: e61366.

Hardisty, A. & Haston, E. (2021). MIDS Level 1: Specification, conformance checklist, mapping template and instructions for use. *Biodiversity Information Science and Standards* 5: e75574.

Harwell Restoration. (2021). Available at: https://www.harwellrestoration.co.uk

Haston, E., Richardson, J.E., Stevens, P.F., Chase, M.W. & Harris, D.J. (2009). The Linear Angiosperm Phylogeny Group (LAPG) III: a linear sequence of the families in APG III. *Bot. J. Linn. Soc.* 161: 128–131.

Hawthorne, W.D. & Jongkind, C.C.H. (2006). *Woody Plants of Western African Forests.* Royal Botanic Gardens, Kew.

Häuser, C.L., Steiner, A., Holstein, J. & Scoble, M.J. (eds) (2005). *Digital Imaging of Biological Type Specimens. A Manual of Best Practice. Results from a study of the European Network for Biodiversity Information.* Stuttgart.

Heberling, J.M. & Isaac, B.L. (2017). Herbarium specimens as exaptations: New uses for old collections. *Amer. J. Bot.* 104: 963–965.

Heberling, J.M. & Isaac, B.L. (2018). iNaturalist as a tool to expand the research value of museum specimens. *Appl. Plant Sci.*, 6, e01193.

Heberling, J.M., Prather, L.A. & Tonsor, S.J. (2019). The changing uses of herbarium data in an era of global change: An overview using automated content analysis. *Bioscience*, 69, 812–822.

Heberling, J.M., Miller, J.T., Noesgaard, D., Weingart, S.B. & Schigel, D. (2021). Data integration enables global biodiversity synthesis. *Proc. Natl. Acad. Sci.*, 118, e2018093118.

Huxley, R., Quaisser, C., Butler, C.R. & Dekker, R.W.R.J. (2021*). Managing Natural Science Collections: A Guide to Strategy, Planning and Resourcing.* Routledge, London.

I

iDigBio (2015). *Specimen Barcode and Labeling Guide.* https://www.idigbio.org/wiki/index.php/Specimen_ Barcode_and_Labeling_Guide

IUCN (2012). *IUCN Red List Categories and Criteria: Version 3.1, Second edition.* IUCN, Gland and Cambridge.

J

James, S.A., Soltis, P.S., Belbin, L., Chapman, A.D., Nelson, G., Paul, D.L. & Collins, M. (2018). Herbarium data: Global biodiversity and societal botanical needs for novel research. *App. Plant Sciences* 6(2): e1024.

Janssen, T. (2006). Moulding method to preserve tree fern trunk surfaces including remarks on the composition of tree fern herbarium specimens. *Fern Gazette* 17(6/8): 351.

Jarzen, D.M. & Nichols, D.J. (1996). Chapter 9: Pollen. In: J. Jansonius & D.C. McGregor (eds), *Palynology: Principles and Applications,* 1: 261–291. American Association of Stratigraphic Palynologists Foundation.

Jarzen, D.M. & Jarzen, S.A. (2006). Collecting pollen and spore samples from herbaria. *Palynology* 30: 111–119.

Jennings, L., Hope, R. & van der Burgt, X. (2018). *Making herbarium specimens.* Technical Information Sheet 15. Available at: http://brahmsonline.kew.org/Content/Projects/msbp/resources/Training/15-making-herbarium-specimens.pdf

Jensen, E. & Buckley, N. (2014). Why people attend science festivals: interests, motivations and self-reported benefits of public engagement with research. *Public Understanding of Science* 23: 557–573.

John, D.M., Whitton, B.A. & Brook, A.J. (eds). (2011). *The Freshwater Algal Flora of the British Isles: an Identification Guide to Freshwater and Terrestrial Algae.* Second Edition. Cambridge University Press, Cambridge.

Julius, A., Tagane, S., Kajita, T. & Utteridge, T.M.A. (2021). *Ardisia pyrotechnica* (Primulaceae-Myrsinoideae), a new species from Borneo. *Phytotaxa* 507: 205–210.

K

Keller, H.A. & Goyder, D.J. (2021). Una nueva especie de *Philibertia* (Apocynaceae) de Bolivia. *Darwiniana, Nueva Serie* 9: 293–298.

King, C. (2022). *The Kew Book of Botanical Illustration.* Search Press in association with Royal Botanic Gardens, Kew.

Kurz, S., Frietag, A., Nyffeler, R., Scheyer, T., Troxler, M., Holz, B., Stauffer, F., Neissekenwirth, F. Neubert, E. Mazenhauer, J., Greeff, M., Baur, H. Liersch, S., Vihelmsen, L. Litman, J., Cibois, A., Hofmann, B., Price & M., Frick, H. (2021). Object Storage. In: H. Frick & M. Greeff (eds), *Handbook on natural history collections management – A collaborative Swiss perspective.* Swiss Academies Communications 16(2).

L

Lendemer, J., Thiers, B., Monfils, A.K., Zaspel, J., Ellwood, E.R., Bentley, A., LeVan, K., Bates, J., Jennings, D., Contreras, D., Lagomarsino, L., Mabee, P., Ford, L.S., Guralnick, R., Gropp, R.E., Revelez, M., Cobb, N., Seltmann, K. & Aime, M.C. (2020). The Extended Specimen Network: A strategy to enhance US biodiversity collections, promote research and education. *Bioscience* 70: 23–30.

Letouzey R. (1986). *Manual of forest botany, tropical Africa.* Vol. 1: *General botany.* CTFT, Nogent-sur-Marne. Also available in French.

Liesner, R. (2017). Field Techniques Used by Missouri Botanical Garden. [Online]. Available at: http://www.mobot.org/MOBOT/molib/fieldtechbook/welcome.shtml. [Accessed: 22/10/2021].

Lundholm, M. (2019) Chapter 8, Functional Planning for Collection Storage. In: L. Elkin and C.A. Norris (eds), *Preventive Conservation: Collection Storage. Society for the Preservation of Natural History Collections.* American Institute for Conservation of Historic and Artistic Works; Smithsonian Institution; The George Washington University Museum Studies Program.

M

MacGregor, N. (2010–2020). *A History of the World in 100 Objects.* BBC Radio 4 – A History of the World in 100 Objects – Downloads available at BBC Radio 4 – A History of the World in 100 Objects

Maekawa, S. & Elert, K. (2003). *The Use of Oxygen-Free Environments in the Control of Museum Insect Pests.* Tools for Conservation. Los Angeles: Getty Conservation Institute. http://hdl.handle.net/10020/gci_pubs/oxygen_free_enviro

Maina, S.M. (2015). Looking, Seeing and Learning: The Role of Design in Developing Exhibition and Display for Museums. *Africa Habitat Review, Journal of The School of the Built Environment* 9: 805–822.

Makos, K.A., Hinkamp, D. & Smith Jr., J.R. (2019). Safety and health issues within storage spaces. In: L. Elkin and C.A. Norris (eds), *Preventive Conservation: Collection Storage. Society for the Preservation of Natural History Collections.* American Institute for Conservation of Historic and Artistic Works; Smithsonian Institution; The George Washington University Museum Studies Program.

Mars, R. (2010–present). *99% Invisible.* Available at 99% Invisible (99percentinvisible.org)

Martin, G.J. (2004). *Ethnobotany: a methods manual.* Earthscan, Abingdon.

Maurin, O., Epitawalage, N., Eiserhardt, W., Forest, F., Fulcher, T. & Baker, W.J. (2017). *A visual guide to collecting plant tissues for DNA.* Royal Botanic Gardens, Kew. Available at: https://www.kew.org/sites/default/files/2019-07/PAFTOL%20sampling%20guide.pdf

Miquel, F.A.W. (1855–1858). *Flora Nederlands Indie Vol. 1* (1), 1–1116.

Miquel, F.A.W. (1867). *Annales Musei Botanici Lugduno-Batavi Vol. 1.*

Moore, B.P., Weatherson, J.C., White, R.D. & Williams, S.L. (2019). Storage Furniture. In: L. Elkin & C.A. Norris (eds), *Preventive Conservation: Collection Storage. Society for the Preservation of Natural History Collections*, pp. 615–639. American Institute for Conservation of Historic and Artistic Works; Smithsonian Institution; The George Washington University Museum Studies Program.

Morton, A. (15/01/2020). *'Dinosaur trees': firefighters save endangered Wollemi pines from NSW bushfires.* The Guardian.

Museum Association. (2014). Disposal Toolkit, Guidelines for Museums. https://media.museumsassociation.org/app/uploads/2020/06/11090021/31032014-disposal-toolkit-8.pdf

Museum of London (2013). *Museum Pests e-learning tool. Introduction to Museum Pests, Section 4, Useful tools, Floor plans and databases.* https://www.museumoflondon.org.uk/Resources/e-learning/introduction-to-museum-pests/s04p04.html

N

National Archives. (2022). *Advice and guidance – The National Archives. Wearing gloves to handle our documents.* Available at: https://cdn.nationalarchives.gov.uk/documents/information-management/what-is-the-policy-on-wearing-gloves-to-handle-documents.pdf

National Museums Scotland. (2022). Advice for Museums. Collections Care Training. Object Handling and Packing Guidelines. https://www.nms.ac.uk/about-us/our-services/training-and-guidance-for-museums/collectionscare-training/object-handling/

National Park Service (NPS) (1999). H. Dust and Gaseous Air Pollution. In: *NPS Museum Handbook, Part I. Museum Collections. Chapter 4: Museum Collections Environment.* Available at: https://www.nps.gov/museum/publications/MHI/CHAPTER4.pdf

NatSCA (Natural Sciences Collections Association). (2013). *Care and Conservation of botanical specimens*. https://www.natsca.org/sites/default/files/publications-full/care-and-conservation-of- botanical-specimens.pdf

Natural History Museum. (2014). *Standards in the Care of Botanical Materials*. Conservation and Collections Care Scratchpad. https://conservation.myspecies.info/node/35

Nelson, G., Paul, D., Riccardi, G. & Mast, A. R. (2012). Five task clusters that enable efficient and effective digitization of biological collections. *ZooKeys* 209: 19–45.

Neumann, D., Carter, J., Simmons, J.E., Crimmen, O. (2022). *Best Practices in the Preservation and Management of Fluid-preserved Biological Collections*. Society for the Preservation of Natural History Collections.

Newton, L.E. (2004). The first herbarium botanist in Nairobi. *J. East African Natural History* 93: 49–55.

Nordling, L. (2021). The damage is total: fire rips through historic South African library and plant collection. *Nature* 592: 672.

P
Paine, C. (ed.) (1992). *Standards in the Museum Care of Biological Collections*. Museums and Galleries Commission, London.

Paton, J.A. (1999). *The Liverwort Flora of the British Isles*. Harley Books, Colchester.

Pearce, T.R., Filer, D. & Van den Berghe, E. (1996). Computerisation of the East African Herbarium: development of a regional plant information service. In: L.J.G. van der Maesen, X.M. van der Burgt & J.M. van Medenbach de Rooy (eds), *The Biodiversity of African Plants*. Springer, Dordrecht.

Pinniger, D. (2008). *Pest Management: a practical guide*. Collections Trust, Cambridge.

Pinniger, D. (2015). *Integrated Pest Management in Cultural Heritage*. Archetype Publications, London.

Pinniger, D. & Lauder, D. (2018). *Pests In Houses Great & Small: identification, prevention, eradication*. English Heritage, London.

Pollan, M. (2001). *The Botany of Desire: a plant's-eye view of the world*. Bloomsbury Publishing, London.

Poole, N. & Dawson, A. (2013). *Spectrum Digital Asset Management*. Collections Trust, England.

Powell, C., Motley, J., Qin, H. & Shaw, J. (2019). A born-digital field-to-database solution for collections based research using collNotes and collBook. *App. Plant Sciences* 7(8): e11284.

PPG I (2016). A community-derived classification for extant lycophytes and ferns. *J. Syst. Evol.* 54: 563–603.

Purewal, V. (2019). Plant Material. In: L. Elkin & C.A. Norris (eds), *Preventive Conservation: collection storage*, pp. 882–883. Society for the Preservation of Natural History Collections; American Institute for Conservation of Historic and Artistic Works; Smithsonian Institution; The George Washington University Museum Studies Program.

R
Rabeler, R.K., Svoboda, H.T., Thiers, B., Prather, L.A., Macklin, J.A., Lagomarsino, L.P., Majure, L.C. & Ferguson, C.J. (2019). Herbarium Practices and Ethics, III. *Syst. Bot.* 44: 7–13.

Rankin, K.B. (1992). *Mounting System for Herbarium Specimens.* Storage Techniques for Art Science & History Collections (stashc.com).

Rieger, T. (2016). *Technical Guidelines for Digitizing Cultural Heritage Materials, Creation of Raster Image Files.* Federal Agencies Digital Guidelines Initiative (FADGI).

Rose, E. & Gardiner, L.M. (2021). Networks of Knowledge: The library and herbarium of John Martyn at Cambridge. *Linnean PuLSe.* Issue 48 July 2021: 6–8.

Rotta, E, Caminha de Carvalho e Beltrami, L. & Zonta, M. (2008). *Manual de Prática de Coleta e Herborização de Material Botânico.* Embrapa Florestas, Colombo.

Royal Botanic Garden Edinburgh. (2017). *Guide to Collecting Herbarium Specimens in the Field.* Royal Botanic Garden, Edinburgh.

Royal Botanic Garden of Edinburgh. (2022). *Care and conservation of herbarium specimens.* Available at: https://www.rbge.org.uk/science-and-conservation/herbarium/specimen-preparation-care/care-and-conservation-of-herbarium-specimens/

Royal Horticultural Society. (2015). *Large Colour Chart* (Sixth Revised Edition). RHS.

S

Salick, J., Konchar, K. & Nesbitt, M. (eds). (2014). *Curating Biocultural Collections: a handbook.* Royal Botanic Gardens, Kew.

Savage, S. (1945). *A Catalogue of the Linnean Herbarium.* Linnean Society, London.

Schrenk, J. (1888). Schweinfurth's method of preserving plants for herbaria. *Bull. Torrey Bot. Club* 15: 292–293.

Simmons, J.E. (2014). *Fluid Preservation: a comprehensive reference.* Rowman & Littlefield Lanham, MD.

Singh, H.B. & Subramaniam, B. (2008). *Field Manual on Herbarium Techniques.* National Institute of Science Communication and Information Resources, New Delhi.

Smith, A. & Smith, R. (2004). *The Moss Flora of Britain and Ireland.* Second Edition. Cambridge University Press, Cambridge.

Söderström, L., Hagborg, A., von Konrat, M., Bartholomew-Began, S., Bell, D., Briscoe, L., Brown, E., Cargill, D. C., Costa, D. P., Crandall-Stotler, B. J., Cooper, E. D., Dauphin, G., Engel, J. J., Feldberg, K., Glenny, D., Gradstein, S. R., He, X., Heinrichs, J., Hentschel, J., Ilkiu-Borges, A. L., Katagiri, T., Konstantinova, N. A., Larraín, J., Long, D. G., Nebel, M., Pócs, T., Puche, F., ReinerDrehwald, E., Renner, M. A., Sass-Gyarmati, A., Schäfer-Verwimp, A., Moragues, J. S., Stotler, R. E., Sukkharak, P., Thiers,B. M., Uribe, J., Váña, J., Villarreal, J. C., Wigginton, M., Zhang, L., & Zhu, R. (2016). World Checklist of Hornworts and Liverworts. *PhytoKeys* 59: 1–828.

Stearn, W.T. (2004). *Botanical Latin: history, grammar, syntax, terminology and vocabulary.* Reprint edition. David & Charles Exeter.

Stevens, P. F. (2001 onwards). Angiosperm Phylogeny Website. Version 14, July 2017 [and more or less continuously updated since]. Available at: http://www.mobot.org/MOBOT/research/APweb/.

Storr, W. (2019). *The Science of Storytelling: why stories make us human, and how to tell them better.* Abrams Press New York.

Strang, T. & Kigawa, R. (2009). *Technical Bulletin 29: Combatting Pests of Cultural Property.* Canadian Conservation Institute, Canadian Heritage, Ottawa.

Struwe, L. & Nitzsche, P. (2020). *Digital Photography for Plant Identification and Problem Diagnosis*. Available at: https://botanydepot.com/2020/07/27/presentation-how-to-photograph-plants-and-more/

Sweeney, P.W., Starly B., Morris P.J., Xu, Y., Jones, A., Radhakrishnan, S., Grassa, C.J. & Davis, C.C. (2018). Large-scale digitization of herbarium specimens: Development and usage of an automated, high-throughput conveyor system. *Taxon* 67: 165–178.

Systematics Association Committee (1962). Terminology of simple symmetrical plane shapes (chart 1). *Taxon* 11: 145–156 & 245–247.

T

Taylor, N.P. (1991). *Melocactus* in Central and South America. *Bradleya* 9: 1–80.

Thiers, B. (Continuously updated). *Index Herbariorum: A global directory of public herbaria and associated staff.* New York Botanical Garden's Virtual Herbarium. Available at: http://sweetgum.nybg.org/science/ih/

Thiers, B. M. (2020). *Herbarium: the quest to preserve and classify the world's plants.* Timber Press, Portland, OR.

Thiers, B. (2022). *The World's Herbaria 2021: A Summary Report Based on Data from Index Herbariorum Issue 6.0*, published February 2022. http://sweetgum.nybg.org/science/wp-content/uploads/2022/02/ The_Worlds_Herbaria_Jan_2022.pdf

Timbrook, J. (2014). Curating ethnographic specimens. In: J. Salick, K. Konchar & M. Nesbitt (eds), *Curating Biocultural Collections: a handbook*, pp. 27–39. Royal Botanic Gardens, Kew.

Townsend, J. (1999). Agents of Deterioration in Collections: Actions and Interactions. In: D.A. Metsger & S.C. Byers (eds), *Managing the Modern Herbarium*, pp. 27–34. Society for the Preservation of Natural History Collections, Royal Ontario Museum.

Tucker, A.O. & Calabrese, L. (2005). *The use and methods of making a herbarium/Plant specimens: An Herb Society of America Guide.* The Herb Society of America, Ohio. herbarium. pub (herbsociety.org).

U

UN Environmental Programme (UNEP). (2005). *ABS Clearing House. [Online]*. Available at: https://absch.cbd.int/. [Accessed 03/09/2021].

UN Environmental Programme (UNEP). (2006). *Convention on Biological Diversity. [Online]*. Available at: www.cbd.int. [Accessed 03/09/2021].

UN Environmental Programme (UNEP). (2021). *The Species+ Website.* Nairobi, Kenya. Compiled by UNEP-WCMC, Cambridge, UK. Available at: www.speciesplus.net.

Utteridge, T. & Bramley, G. (eds) (2015). *Tropical Plant Families Identification Handbook.* Royal Botanic Gardens, Kew.

V

van Balgooy, M.M.J., Low, Y.W. & Wong, K.M. (2015). *Spot Characters for the Identification of Malesian Seed Plants.* Natural History Publications, Borneo.

van Dormolen, H. (2012). *Metamorfoze Preservation Imaging Guidelines*. National Archives of the Nederlands.

van Steenis-Kruseman, M.J. & van Steenis, C.G.G.J. (1950). Malaysian plant collectors and collections: being a Cyclopaedia of botanical exploration in Malaysia and a guide to the concerned literature up to the year 1950. *Flora Malesiana – Series I, Spermatophyta* 1: 2–639.

Victor, J.E., Koekemoer, M., Fish, L., Smithies, S.J. & Mössmer, M. (2004). *Herbarium Essentials: the southern African herbarium user manual.* Southern African Botanical Diversity Network Report, no. 25. National Botanical Institute, Pretoria.

W

Walters, S. M. (1981). *The shaping of Cambridge botany: a short history of whole-plant botany in Cambridge from the time of Ray into the present century.* Cambridge: Cambridge University Press, pp. 30–35.

Way, M. & Gold, K. (2014). *Assessing a population for seed collection.* Technical Information Sheet 2. Available at: http://brahmsonline.kew.org/Content/Projects/msbp/resources/Training/02-Assessing-population.pdf

Webster, M.S. (ed.) (2017). *The Extended Specimen: Emerging Frontiers in Collections-Based Ornithological Research.* CRC Press, Boca Raton, FL.

Western Australian Museum (WAM). (2017). *Care Manual.* Western Australian Museum. Available at: https://manual.museum.wa.gov.au/handling-objects

Whitmore, T.C. (1972). The description of a tree. Chapter 2 in *Tree Flora of Malaya*, Vol. 1. Longman Group, London.

Wilkie, P. (2013). The collection and storage of plant material for DNA extraction: The Teabag method. *Gardens' Bulletin Singapore* 65(2): 231–234.

Wilkinson, M.D., Dumontier, M., Aalbersberg, I.J., Appleton, G., Axton, M., Baak, A., Blomberg, N., Boiten, J.-W., Bonino da Silva Santos, L., Bourne, P.E., Bouwman, J., Brookes, A.J., Clark, T., Crosas, M., Dillo, I., Dumon, O., Edmunds, S., Evelo, C.T., Finkers, R., Gonzalez-Beltran, A., Gray, A.J.G., Groth, P., Goble, C., Grethe, J.S., Heringa, J.,'t Hoen, P.A.C., Hooft, R., Kuhn, T., Kok, R., Kok, J., Lusher, S.J., Martone, M.E., Mons, A., Packer, A.L., Persson, B., Rocca-Serra, P., Roos, M., van Schaik, R., Sansone, S.-A., Schultes, E., Sengstag, T., Slater, T., Strawn, G., Swertz, M.A., Thompson, M., van der Lei, J., van Mulligen, E., Velterop, J., Waagmeester, A., Wittenburg, P., Wolstencroft, K., Zhao,J. & Mons, B. (2016). The FAIR Guiding Principles for scientific data management and stewardship. *Scientific Data* 3: article number 160018. https://doi.org/10.1038/sdata.2016.18

Womersley, J.S. (1981). *Plant collecting and herbarium development: A manual.* FAO Plant Production and Protection Paper, No. 33. Food and Agriculture Organisation of the United Nations, Rome.

Wondafrash, M., Wingfield, M.J., Wilson, J.R., Hurley, B.P., Slippers, B. & Paap, T. (2021). Botanical gardens as key resources and hazards for biosecurity. *Biodiversity and Conservation* 30(7): 1929–1946.

Woodruff, L. (2008). *How to Pack Herbarium Specimens for a Loan.* How_To_3.pdf (spnhc. org), University of Texas at Austin.

Y

Yockteng, R., Almeida, A.M.R., Yee, S., Andre, T., Hill, C. & Specht, C.D. (2013). A method for extracting high-quality RNA from diverse plants for next-generation sequencing and gene expression analyses. *Appl. Plant Sci.* 1(12): apps.1300070.

Z

Zerpa-Catanho, D., Zhang, X., Song, J., Hernandez, A.G. & Ming, R. (2021). Ultra-long DNA molecule isolation from plant nuclei for ultra-long read genome sequencing. *STAR Protoc.* 2(1): 100343.

IMAGE CREDITS

The editors and contributors are extremely grateful to the following photographers for allowing use of their images. Photographers are listed below according to the chapter in which their images appear and the image numbers on the relevant page. Where not listed below, the images are in the public domain or the copyright is owned by the Royal Botanic Gardens, Kew.

Introduction

P2: Trustees of the Natural History Museum, London (2), The Royal Horticultural Society (RHS)/Adrian Green (3). P5: Manuel Lujan (1, 3–5), Carrie Kiel (2). P9: Timothy Utteridge (1–4).

Collecting for the herbarium

P10: Marie Briggs. P13: William J. Baker. P15: William Milliken. P18–19: Alan Paton. P20–21: Nina M.J. Davies (1–3). P23: Royal Botanic Garden Edinburgh (RBGE). P25: Nomentsoa Randriamamonjy. P32–33: M. Briggs (1a–c,2,7), Liam Trethowan (3,4), Gemma Bramley (5), Xander van der Burgt (6), Thomas Heller (8), Kunanon Daonurai (9). P35: Naiyana Tetsana (10,12). Anna Haigh (13). Tiina Sarkinen (14). Jean-Michel Touche (15). P36: K. Daonurai (16). M. Briggs (17,19). Jo Osborne (18). Victor Simbiak (20). P.39: Joanna Wilbraham (1–6, 8), Alex Monro (7). P41: Juan Viruel (1–3), Lauren Gardiner (4–6), W.J. Baker (7). P42–43: Nicola Biggs (1,3), Nigel Taylor (2), Thomas Heller (5). P45: N.M.J. Davies (10). P47: M. Briggs (1, 2), Mark Watson (3, 4), X. van der Burgt (5–7). P48: A. Haigh (8, 9), X. van der Burgt (10), K. Daonurai (11–13). P51: K. Daonurai (1–4). P55: Denise Molmou (1,2). P57: Wittawat Kiewbang (1), K. Daonurai (2,3). P59: Simone Soares da Silva (1), Sérgio Bianchini (2), Mayara Pastore (3). P60: Olivier Maurin (1–5). P62: Theerawat Thananthaisong (1,2). P63: X. van der Burgt (1–3). P66: Daniel Cahen (1,2). P68: D. Cahen (1–4). P70: Kennedy Matheka (1–5).

Herbarium techniques

P87: Carnegie Museum of Natural History (1), Mid-Atlantic Herbaria Consortium (2), iNaturalist.org (3), TRY-db.org (4-upper), Mason Heberling (4-lower, 7 top left, 7 bottom right, 8-lower, 9), Biodiversity Heritage Library/Smithsonian Institution (7 top right), Allison Heberling (7 bottom left), Abby Yancy (8-upper), Global Biodiversity Information Facility/OpenStreetMap contributors (10). P101: Martin Cheek (1–4). P102: X. van der Burgt (1), David Goyder (2), Shuichiro Tagane (3), Fernando Zuloaga (4). P105: © RHS Wisley (1,3). P120: X. van der Burgt (1–2). P131: K. Fujikawa (5). P139: T. Utteridge (3), N.M.J. Davies (4). P140: T. Utteridge (2). P148: © National Museums Liverpool, World Museum (1–5). P151: © National Museums Liverpool, World Museum (1–6). P152–3: © National Museums Liverpool, World Museum (1–6). P155: Clare Drinkell (2). P157: C. Drinkell (4). P167: M. Briggs (2), Marie-Hélène Weech (3,5), Marcella Corcoran (4). P169: T. Utteridge (5). P179: Federico Rossi (1a+b), Cícero Rodrigues (2–4, 6), Luis Fernando (5).

Building and environment

P183: N.M.J. Davies (4). P187: C. Drinkell (2). P191: C. Drinkell (1). P195: M. Briggs (2), DBP Entomology (3), C. Drinkell (7), P. Viscardi (8). P197: UK Crown Copyright – courtesy of Fera and Collections Trust (1, 3, 4, 6–9, 11–15), DBP Entomology (2), X. van der Burgt (5), ©Malcolm Storey www.bioimages.org.uk (10). P199: UK Crown Copyright – courtesy of Fera and Collections Trust (1, 4), X. van der Burgt (2), ©Malcolm Storey www.bioimages.org.uk (3), DBP Entomology (5), Alison Moore (6), Alan Outen (7), © Historic England. English Heritage Trust (8). P200: Yvette Harvey (1, 2, 4), Elizabeth Woodgyer (3). P210: Collection Coordinator/Herbarium Bogoriense (all images).

The herbarium in the wider context

P212: © Paul DuBois. P215: N.M.J. Davies (1), National Herbarium, South African National Biodiversity Institute (2). P217: © NYBG (1). P219: ©Royal Botanic Garden Edinburgh (2-3). P220: C. Drinkell. P222–223: CSIRO (1, 2, 5), Brendan Lepschi (3, 4). P226: Jenny Murray (1). P229: Gemma Bramley and Lesley Walsingham (1), Melissa Bavington (2). P231: © P. DuBois, Yvette Harvey © RHS (2–4). P233: C. Drinkell (1–3), X. van der Burgt (4). P235: N.M.J. Davies (2). P237: Cambridge University Herbarium (1–4, 6), Amber Horning (5). P239: Isabel Larridon (1), Martin Cheek (2), Charlotte Couch (3–4), HNG (5). P241: Kochi Prefectural Makino Memorial Foundation, Inc. (1–6). P243: Emily Magnaghi (1–5), Nathalie Nagalingum (6). P244–245: © Royal Botanic Garden Edinburgh (1–6). P247: Singapore Botanic Gardens (1–4). P248: © NYBG (1, 2).

Cultivated specimen from the Garden of
Mr. EDWARD LEEDS, Received May 1876.

Aglaia

June 1876

Tulipa gesneriana   Longford Bridge

INDEX

Numbers in **bold** indicate the presence of a relevant image.